# MUSEOLOGY

## TEXAS TECH UNIVERSITY

### Guidelines for Managing
### Bird Collections

*Paisley S. Cato*

No. 7

April 1986

# TEXAS TECH UNIVERSITY

Lauro F. Cavazos, President

Museology
Texas Tech University
No. 7, 78 pp.
25 April 1986

Museology, Texas Tech University, is a Texas Tech Press serial publication of The Museum. Titles are numbered serially, paged separately, and published on an irregular basis under the auspices of the Vice President for Academic Affairs and Research and in cooperation with the International Center for Arid and Semi-Arid Land Studies. Copies may be purchased from Texas Tech Press, Sales Office, Texas Tech University, Lubbock, Texas 79409, U.S.A.

ISSN 0196-0237
ISBN 0-89672-130-2

Texas Tech Press, Lubbock, Texas

1986

# CONTENTS

# GUIDELINES FOR MANAGING BIRD COLLECTIONS

PAISLEY S. CATO

Surveys by the American Ornithologists' Union (Banks *et al.*, 1973; Clench *et al.*, 1976) report the existence of more than four million bird study skins in collections in Canada and the United States. The importance of these collections both for research and other phases of museum operations have been discussed by many authors (Mayr and Goodwin, 1956; Parkes, 1963; Zusi, 1969; Allen, 1974; Banks, 1979; Keast, 1973; Ricklefs, 1980; Barlow and Flood, 1983). Both large and small collections are valuable reservoirs of information. However, if collections are to "yield information, contribute knowledge or provide stimuli for aesthetic responses, they must be available for study and to view" (Force, 1975). As with other forms of documentation, specimens are useless if they are not accessible for further study.

The goal of collection management is two-fold: 1) to ensure the preservation of specimens and their data, and 2) to permit retrieval of both the specimens and their data. Collection management in the natural sciences requires a knowledge not only of taxonomy, but of information handling systems, archival techniques, and certain procedures unique to museum collections. Specimens and their data must be protected from destructive agents such as extreme temperatures and humidity, light, dust, and insect pests. They must be organized so as to permit efficient retrieval. Care must be taken to maintain a high level of accuracy in the data accompanying the specimens. Organizations such as the American Ornithologists' Union (King and Bock, 1978) and the Association of Systematics Collections (Lee *et al.*, 1982) have encouraged and adopted professional standards for the management of scientific collections.

Traditionally, curators, collection managers, and collection assistants have been trained on the job. Written guidelines for managing bird collections have been scattered and available only in journal articles, each dealing with just a small facet of the entire task. This guide is an attempt to provide in a single source an overview of the procedures and techniques required to manage a bird collection. The information presented here reflects the practices that are in use in several of the larger North American bird collections.

In 1983, 33 curators of bird collections, which contain approximately 4.3 million bird specimens, were surveyed by the author concerning some of their collection management techniques. Twenty-nine of the 33 institutions responded to a questionnaire that covered the topics of acquisitions, specimen preparation, registration, storage, maintenance, and collection use. These collections are listed in Appendix I. Results from the questionnaires indicated that in each of these phases of collection management, no one single procedure is used exclusively. Instead, based on the overall goal of collection management, general procedures have been adapted to fit the needs of the individual institution.

### COLLECTION MANAGEMENT POLICY AND PROCEDURE MANUALS

A manual of policies and procedures serves as a vehicle whereby an institution can limit the scope of its collections, as well as establish standards for the management and use of specimens in its collection. It sets forth the purposes and goals of the institution, and explains how these are interpreted in its collections operations (Malaro, 1979). This manual may also act as a valuable tool for training staff and volunteers.

Each institution or collection should develop its own set of policies and procedures to fit its own needs and circumstances. Larger institutions frequently have a general policy for the entire museum, and secondary departmental procedures for specific collections (Field Museum of Natural History, 1976; Denver Museum of Natural History, 1978; National Museums of Canada, 1983). A number of topics should be considered when developing a set of guidelines:

1. Purpose of the institution or collection and proposed scope of the collection
2. Acquisitions: what is to be acquired, how, by whom, under what conditions, limitations, ethical and legal constraints
3. Specimen documentation: accessioning, cataloging, field journals, and the like
4. Information retrieval: manual and computer
5. Storage and conservation: organization of specimens, inventory, fumigation, environmental controls, insurance, and the like
6. Access to collection: by whom, under what conditions, and loan policies
7. Deaccessions: exchanges, disposal, how, why, by whom, ethical, and legal constraints

Examples of policies and procedures currently used in bird collections are included in the following sections of this guide. Additional information is available from a variety of sources. Many of the procedures used for other zoological collections are applicable for ornithological collections. These are described in Zweifel (1966), Wake (1975), Williams *et al.* (1977), and Fink *et al.* (1978). Two professional museum organizations, The American Association of Museums and the American Association for State and Local History, have published a number of articles and books pertaining to the various aspects of collection management. The Association of Systematic Collections is a particularly good source for information on Federal collecting regulations and problems of universal concern to systematics collections (Humphrey, 1972; Trueb and Edwards, 1978; Association of Systematics Collections, 1979, 1981). Addresses for these and other relevant organizations are listed in Appendix II.

## ETHICS AND LEGALITIES

### *Ethical Standards*

Collection management policies and procedures should be based on sound ethical and legal principles. The American Ornithologists' Union (1975), the American Association of Museums (1978), and the International Council on Museums (1970) are among the professional organizations that have prepared codes of ethics for institutions and their personnel with responsibilities for collections. Although a collection management manual cannot exhaustively cover every situation, an attempt should be made to recognize and consider the most common situations. Guidelines included in a manual should delineate for personnel the professional standards for these situations.

A particular area of concern should be specimen collection. All personnel involved in collecting birds, eggs, and nests should be aware of and follow all legal regulations. They should have all permits required by the federal and state laws of the United States and any other country in which collection takes place. Aside from the possibility of being punished by law if legalities are not strictly observed, bird collectors need to consider the impact of their actions on public opinion. Killing birds is a delicate subject for many people, and an indiscriminate collector can prompt unjust criticism of all ornithologists and museum professionals (American Ornithologists' Union, 1975). Misuse of collecting privileges can anger land owners who may refuse permission to collect additional specimens on their property.

The "Code of Ethics for Collectors and Capturers of Wild Birds" (American Ornithologists' Union, 1975) summarized the ethical considerations involved in specimen collection:

1. The privileges of a collecting or capturing permit shall be used only to obtain specimens for justifiable scientific or educational purposes
2. Collect or capture specimens only from those populations or species that can sustain the loss of individuals
3. Collect or capture only those specimens that are deemed necessary and that can be properly cared for or prepared
4. Exercise the greatest care in recording accurately the maximum amount of relevant data for all specimens obtained
5. If live birds are collected, maintain them under humane conditions with high standards of health and sanitation
6. Collect with the aim of making available all relevant data obtained from specimens, either through publication or by giving others access to the data
7. Abide by all stated regulations, including the use of authorized permits to collect, capture, import, export, and transship specimens
8. Notify the appropriate local authorities of plans to collect or capture birds in areas under their jurisdiction

9. Identify yourself and your purposes to those who may witness your collecting or capturing in order to inform them of the validity of your activities

10. Be as judicious and humane as possible in collecting and capturing activities, taking care to respect the rights, interests, and feelings of others

11. Regard the privilege to collect or capture birds as a trust in the pursuit of science; it should never be flaunted

Museum staff members should be aware that some forms of personal collecting can create other ethical problems. *Museum Ethics* (American Association of Museums, 1978) suggests that "No employee may compete with his institution in any personal collecting activity." Museum employees with the proper state hunting licenses may collect game birds in season and other nonprotected species for personal use or profit without fear of prosecution. However, in some circumstances, this might create a conflict of interest, and such activities should be disclosed to a supervisor or the museum administration (Ullberg and Ullberg, 1974; American Association of Museums, 1978).

Personal collecting does not commonly create a conflict of interest for ornithologists due to the strict laws regarding the possession of ornithological materials. It is illegal in the United States to possess private holdings of protected species collected after 1970, and few collectors are willing to risk prosecution to maintain personal collections. Federal and state collecting permits that allow legal collection of numerous specimens also specify that the specimens be deposited in educational or scientific institutions.

In addition to ethical and legal collecting procedures, employees associated with collections should recognize the need to provide adequate care for the specimens. Specimens are wasted if they are not protected from destructive agents. Museum personnel are responsible for providing an adequate physical environment, for ensuring proper documentation for all specimens, and for allowing only responsible persons to have access to the specimens.

Even with the best of care, it is obvious that many specimens wear out or lose their value and therefore need either to be deaccessioned or disposed of. A committee should be formed to oversee this process so that no one curator or director can exercise absolute power over what is deaccessioned. Some provision should also be made if in the future the collection becomes "orphaned" without staff or financial resources to care for the specimens.

Museum staff members are ethically obligated to see that their deportment does not reflect negatively on their institution. Any museum-related actions, even those beyond specimen collection and care, may reflect on that museum or be attributed to it (American Association of Museums, 1978).

## Legal Requirements

### Collection of Specimens

Many state and Federal laws govern the collection, possession, and transport of birds, eggs, and nests for educational and scientific purposes. Laws frequently change, as do permit requirements, and no attempt will be made in this paper to give an exhaustive listing or interpretation of those laws. Instead, the primary laws and sources for additional information will be presented. It is the responsibility of each professional to contact the proper authorities in order to ensure compliance with all laws.

The Federal regulations pertaining to all United States wildlife are contained in the *Code of Federal Regulations 50* published by the United States Government Printing Office. The categories of birds governed by the different types of legislation are as follows (American Ornithologists' Union, 1975):

1. Migratory game birds including Anatidae, Gruidae, Rallidae, Phalaropodidae, Recurvirostridae, Scolopacidae, Charadridae, Haematopodidae, Jacanidae, and Columbidae
2. Resident game birds including members of the order Galliformes
3. Migratory nongame birds
4. Migratory insectivorous birds as defined by the Migratory Bird Treaty between the United States and Canada
5. Endangered species as based on various statutes including the Endangered Species Act of 1973
6. Injurious wildlife whose importation is controlled based on the Lacey Act
7. Poultry that is subject to quarantine and import laws
8. Game birds
9. Eagles, which are protected in the United States under the Eagle Protection Act as amended
10. Birds for research purposes
11. Commercial birds

Lists of species included in these categories may be obtained by writing the U.S. Fish and Wildlife Service, Washington, D.C. 20240, and the Canadian Wildlife Service, Ottawa K1A 0H3, Ontario, Canada. Other comprehensive lists of Federally controlled species are the *Directory of Federally Controlled Species* (Berger *et al.*, 1979) and the *Index to U.S. Federal Wildlife Regulations* (Berger and Phillips, 1977).

The principle Federal laws governing bird species are:

1. Bald and Golden Eagle Protection Act (as amended) P.L. 92-535; 16 U.S.C. 668-668D
2. Endangered Species Act of 1973 (as amended) 16 U.S.C. 1531-1543
3. Lacey Act of 1900 (as amended) 16 U.S.C. 702-2; 18 U.S.C. 42-44
4. Migratory Bird Treaty of 1918; 16 U.S.C. 703

For explanations and interpretations of these and other laws pertaining to systematics collections, see American Ornithologists' Union (1975), Edwards and Grotta (1976), Hart (1978), Phelan (1982), and Bean (1983).

In addition to the Federal legislation protecting bird species, each state has its own laws. The protection status of many species as well as application and reporting requirements for permits vary among the states; two comprehensive listings of state controlled species and requirements are available (McGaugh and Genoways, 1976; Berger and Phillips, 1981).

Persons involved in collecting birds, nests, or eggs, and salvaging dead birds must have both a Federal Scientific Collecting Permit and the state permit from the state in which collecting or salvaging will be done. Applications for a Federal Scientific Collecting Permit may be obtained by writing to the Special Agent in Charge, U.S. Fish and Wildlife Service, for the district involved (see Appendix II for those addresses).

Federal permits are issued to individuals, not to institutions. Applicants for a Federal permit are required to outline the species and number of migratory birds or their parts, eggs, and nests to be taken. In addition, applicants must outline proposed collecting localities, justify the permit by including a research project plan, and list the name and address of the institution to which all specimens will be donated. Specimens taken under a permit must generally be transferred to that institution within 60 days following the expiration date of the permit. A report of specimens taken must be filed annually.

Permit holders must have the permit in their possession when collecting; subpermittees are not allowed to kill birds except in the presence of the principle permit holder. Subpermittees may, however, salvage dead birds.

Special permits from the U.S. Fish and Wildlife Service are required for collecting either endangered species or eagles. Special permits also are required from the chief ranger or superintendent of any national park, national forest, national wilderness area, and the like, in order to collect or salvage specimens in those areas.

### Transportation of Specimens

Federal regulations state that protected birds, eggs, and nests may not be shipped across state lines in the United States unless marked plainly with the names and addresses of the shipper and consignee, and with an accurate statement of the contents of the package. In addition, personnel should be aware of any state laws regulating the shipment of bird specimens. The interstate shipment of endangered species taken after December, 1973, requires a permit issued by the Director of the U.S. Fish and Wildlife Service (American Ornithologists' Union, 1975).

The importation of birds, eggs, and nests into the United States is governed by many regulations involving the U.S. Fish and Wildlife Service, the U.S. Department of Agriculture, and the Bureau of Customs. For specimens to be imported, a complete Declaration of Importation of Fish or Wildlife (USFWS Standard Form 3-177) must be filed with the District

Director of Customs at the port of entry (American Ornithologists' Union, 1975; Phelan, 1982). If laws or regulations of the country of origin restrict either the possession or transport of wildlife, the importer may be required to produce documentation, such as permits from the foreign country, to show that no laws have been violated (Phelan, 1982).

Importation of specimens collected outside the United States are regulated by both the Lacey Act and the Convention on International Trade in Endangered Species of Wild Fauna and Flora (CITES). According to the Lacey Act, birds may not be imported into the United States contrary to the laws of the country of origin. Foreign documentation must be provided by the importer to prove the legality of the transaction (American Ornithologists' Union, 1975). Under the CITES treaty, nearly 80 nations have agreed to limit or prohibit the import, export or reexport of a large number of threatened animal and plant species. Lists of the animals regulated by CITES are available from U.S. Fish and Wildlife Service law enforcement officials.

All imported specimens must enter the United States at designated ports-of-entry in containers that are clearly marked with the name and address of the shipper and consignee, and a complete description of the contents, including the species and numbers of each.

International exchanges between institutions of specimens for research purposes are also regulated by CITES. A general permit certifying scientific exchanges may be issued for a two-year period eliminating the need for a separate permit for each exchange.

The importation of live birds (except from Canada or Mexico) requires that an import permit be obtained from the U.S. Department of Agriculture. This agency regulates the manner of care of live animals under the Animal Welfare Act of 1970 (American Ornithologists' Union, 1975; Phelan, 1982).

Endangered species may be imported only as authorized by the Director of the U.S. Fish and Wildlife Service either for scientific purposes or propagation of the species. This requires an application (USFWS Standard Form 3-200) accompanied by a detailed and acceptable justification for a permit to import such species as stated in the Endangered Species Act of 1973 (American Ornithologists' Union, 1975). If the birds are alive, the importer must also 1) report to the Director of the U.S. Fish and Wildlife Service on the importation within 10 days of receipt of the permit, 2) immediately report any deaths and escapes of birds to the Director of the Division of Law Enforcement of the U.S. Fish and Wildlife Service, and 3) retain any carcasses of birds that either die or are killed in such a way as not to impair their use as scientific specimens.

The importation of live birds and eggs defined as injurious under the terms of the Lacey Act requires a permit from the U.S. Fish and Wildlife Service obtainable from the Special Agent in Charge of the Law Enforcement District in which the importer resides. Dead birds, canaries, and psittacine birds are excluded from the provisions of the Lacey Act (American Ornithologists' Union, 1975).

## ACQUISITIONS

### Acquisition Policies

Should an institution accept every specimen that it is offered? Probably not. Practical limitations exist as a result of finite storage space and personnel time. Each institution should determine and limit the scope of its collections in order to maximize the value of its collections, to ensure the most efficient use of the specimens, and to avoid unnecessary collection maintenance costs. It should be emphasized that although very few specimens are useless, the value of many specimens will be greatly enhanced if they are deposited in the appropriate collection.

Fritts (1976) discussed the importance of a written policy to clarify and delineate the institution's policy. Written policies and procedures ensure uniform compliance by both museum staff and collection users. Fifty per cent of the collections surveyed in 1983 reported having a set of written institutional acquisition policies. Several institutions have published their acquisitions and collections policies (American Museum of Natural History, 1974; Nicholson, 1974, 1975; Denver Museum of Natural History, 1978; National Museums of Canada, 1983).

The acquisition policy generally reflects the museum's and collection's purposes. Acquisitions for a small, regional museum may be primarily oriented for hands-on educational use, and generally consist of salvaged specimens. A much larger, more diverse museum may focus on the research emphasis of the collections. For example, the Field Museum of Natural History (1976) listed three priorities in its acquisition policy:

1. Strengthen collection areas in which the museum has a current specialization and recognized historical interest, especially when these areas are threatened irreversibly by the activities of man
2. Broaden the comparative base of our established collection areas
3. Obtain collections of a general nature which are within the broad interests of the museum

Once the scope of the collections is defined, the institution should determine the procedure for deciding what will be accepted, who will have the authority to accept specimens, and how the specimens will actually become property of the institution. It might be desirable to list in the policy manual the criteria which should be considered before a specimen is accepted. The following questions illustrate some points to be considered:

1. Is the specimen relevant to the purpose of the museum?
2. What is its physical condition and has it been adequately prepared? Are there signs of insect infestation?
3. Do the data accompanying the specimen meet the minimal professional standards?
4. What is the potential use of the specimen: research, teaching, and exhibits?
5. Does the institution already have an extensive series of this species over a geographic and temporal range?

6. Can the institution properly care for the specimen, and is it the most appropriate repository for the specimen?
7. Was the specimen legally acquired? Are copies of the permits available?
8. Is the specimen being offered with any restrictions or conditions?

The potential use of a specimen frequently determines how it is prepared and stored, and specimens should be accepted with a specific use in mind. Specimens might become part of a research collection, a teaching collection, or a mounted bird collection. Subdividing the specimens according to their intended use helps to preserve the integrity of the research specimens, and to facilitate access to all specimens.

The research collection should include only those specimens which have potential research value. Most are prepared as study skins, but many institutions also maintain skeletal preparations, fluid-preserved specimens, eggs, and nests as part of the research collection. The numbers of specimens prepared as skeletons and fluid-preserved specimens have increased greatly over the last 10 years in response to the realization of the scientific value of these types of specimens. A world-wide inventory of avian anatomical specimens has recently been initiated by the committee on collections of the American Ornithologists' Union (Zusi *et al.*, 1982; Wood *et al.*, 1982*a*, 1982*b*; Jenkinson and Wood, 1985) to document current resources and to encourage the further collection of those materials. Stomach contents and frozen tissue collections may also be part of the research collection.

Teaching collections are used primarily by educators and artists. Specimens in these collections generally have either little or no research value; thus they may be handled freely. Seventy per cent of the collections surveyed had separate teaching collections, varying in size from a "few" to 6,700 specimens. Specimens in those collections are prepared in any of the possible forms, from study skins to mounted specimens.

Mounted birds are usually maintained separately because of their inconsistent shapes and sizes. Lack of data frequently prohibits the inclusion of mounts in research collections. Of the collections surveyed, none reported having data on more than 10 per cent of their mounted specimens. Many mounted specimens cannot be handled freely because of their fragility. These specimens are, however, of particular importance as exhibit specimens and reference material for artists.

The curator responsible for the collection initially approves all acquisitions. Many institutions establish a policy whereby acquisitions of exceptional value, or requiring special facilities, should also be approved by the museum's collections committee, director, or Board of Trustees (Denver Museum of Natural History, 1978; National Museums of Canada, 1983).

If specimens need to be appraised for tax purposes, the appraisal is the responsibility of the donor. The Internal Revenue Service does not recognize as valid appraisals performed by staff of the recipient museum. This policy functions to avoid possible conflicts of interest. Although museum staff should not perform appraisals, they may act in an advisory capacity to help donors obtain accurate appraisals from outside sources.

*Sources of Acquisitions*

Specimens may be acquired as the result of salvage activities, field work, donations, purchases, and exchanges. The source of acquisition will generally determine the type of documentation which will accompany the specimens and that which should be prepared by the staff of the collection receiving the specimens.

Approximately 41 per cent of the annual growth of bird collections can be attributed to salvage activities (King and Bock, 1978). Birds killed when hitting transmission towers, power lines, picture windows, and cars are picked up by researchers and the general public for donation to institutions. Those specimens become property of the institution under the conditions of Federal salvage permits. Some institutions maintain their salvage specimens separately from other research specimens.

When accompanied by data, salvaged specimens may serve as valuable additions to the data base available for research projects. A study conducted by the Tall Timbers Research Station, Leon County, Florida, was based on bird kills over 25 years from a transmission tower. The salvaged specimens have provided excellent data for studies on pesticide residues, genetic relationships as revealed by protein comparisons, energetics, fat metabolism, and migration patterns (Avise and Crawford, 1981).

Lack of data can be a major problem with salvaged specimens when the specimens are collected by members of the general public. It would be to the collection's best interests to develop some simple procedures for dealing with those specimens, and for educating public donors of the need for accurate data. Some institutions rely on their donation forms to document salvage specimens. Wildlife law enforcement officers should also be notified when raptors and other protected species are obtained in this manner.

Zoo mortalities are included in the salvage figures. These specimens rarely have either complete or accurate data, and are therefore of limited value for research. However, they prove useful for both exhibits and teaching purposes. Zoos frequently provide an institution with exotic species that cannot otherwise be obtained. These specimens also should be documented as completely as possible, and one method would be to develop a "zoo specimen record" such as the one used at the Denver Museum of Natural History (Fig. 1).

The second largest source of acquisitions comes from staff and student collecting, providing 35 per cent of the specimens acquired annually (Barlow *et al.*, 1977). Of the 33 collections surveyed by the author, nine reported that staff acquisitions accounted for 60 to 95 per cent of their collection's annual growth.

Under the conditions stated in state and Federal permits, specimens collected in the field by staff members immediately become property of the institution, and a donation form is not necessary to transfer ownership of the specimens to the institution. It is the responsibility of the collectors either to prepare the specimens, or to have them prepared, and to maintain accurate field notes documenting collection activities. The original field notes should remain at the institution with the specimens, as should collecting permits.

Donations comprise 14 per cent of annual acquisitions (Barlow *et al.*, 1977). These range in form from research specimens to mounted specimens for exhibits and to skins without data for teaching purposes. All possible forms of documentation should be obtained at the time of donation. Sixty-two per cent of the collections surveyed by the author used either a donation form or an accession form to document incoming material. Approximately one-third either used both, or a combined form (Fig. 2).

A donation form should be completed for all specimens given to a collection in order to serve as a transfer of ownership. The donation form may also fulfill record keeping requirements for the Internal Revenue Service for donated specimens being used as charitable contributions. To fulfill both purposes, the form should include the following:

1. Name and address of donor
2. Thorough description of the item(s)
3. Date of the transaction
4. Manner and approximate date the donor acquired the item(s)
5. Circumstances under which the donor came into possession of the item(s) if he/she is not the owner or collector
6. List of material documenting proof of ownership, and relevant information about the donated item(s)
7. History of the item(s) and any other significant information that would contribute to its value or authenticity
8. Original cost or appraisal of the item(s), if applicable
9. Cost, fair market value, or appraised value of the items at the time of donation
10. Names and addresses of appraisers
11. Factors on which the valuation was based
12. Location of the item(s) in the institution

It is in the museum's best interests to include a signed statement that the donor is the lawful owner of the item(s), and that the gift is unconditional. It is generally best not to accept restricted gifts. A representative of the museum also should sign the donation form to indicate acceptance of the gift. Both the donor and the museum should retain copies of the form, and a personal letter of thanks from the curator would emphasize the institution's appreciation for the gift.

Purchases and exchanges provide the remaining acquisitions. Financial restraints usually limit purchases, but exchanges provide an important method of filling gaps in a collection. Exchanges should be mutually beneficial but are not always completed on a one-to-one specimen basis. The relative abundance of the species and its representation in collections also are taken into account. Exchanges are sometimes left "open," a

DEPARTMENT OF ZOOLOGICAL COLLECTIONS
DENVER MUSEUM OF NATURAL HISTORY
CITY PARK, DENVER, CO  80205

ZOO SPECIMEN CHECKLIST

SPECIMEN _____ ____        SEX _____  AGE _____

DONATION DATE_____        DATE OF DEATH _____

DONOR ZOO _____        DONOR ZOO'S SPECIMEN # _____

SPECIMEN INFORMATION:
ORIGINAL DATE AND SOURCE OF ACQUISITION BY DONOR ZOO _____

PLACE OF ORIGIN (NATIVE LOCALITY) _____

TYPE OF PREPARATION     ___SKIN        ___SKELETON        ___OTHER

WEIGHT _____

TOTAL LENGTH (approximate) _____        HEIGHT (approximate)_____

Please include a copy of exact measurements if available.

CAUSE OF DEATH

Are photographs available? _____

REMARKS

When feasible, a cosmetic necropsy is preferred.

Fig. 1.—Zoo Specimen Record used at the Denver Museum of Natural History (original size: 8½ by 11 inches).

**DENVER MUSEUM OF NATURAL HISTORY**
CITY PARK
DENVER, COLORADO 80205
## DONATION RECORD

It is with great appreciation that the Denver Museum of Natural History accepts this gift. It will be utilized in the best interest of the Museum to enhance the collections and for the education and enjoyment of others.

Donations become the full legal property of the Denver Museum of Natural History and will not be accepted if accompanied by restrictions or qualifications of any type or manner.

By affixing (his/her) signature to this document, the donor acknowledges that the gift is unconditional and that (he/she) has the legal capacity to convey the property to the Denver Museum of Natural History.

DONOR'S NAME_____ Donation Date _____

ADDRESS _____ Department _____

_____ Received by _____

PHONE _____ DMNH Accession # _____

1. DESCRIPTION OF DONATION _____

_____

_____

_____

2. COST, DATE AND MANNER OF ACQUISITION BY DONOR _____

_____

3. PROOF OF AUTHENTICITY AND NAMES OF APPRAISERS _____

_____

4. FACTORS ON WHICH VALUATION WAS BASED _____

_____

5. IF NOT AN OWNER, STATE UNDER WHICH CIRCUMSTANCES YOU CAME INTO POSSESSION OF THE OBJECT(S)

_____

_____

_____          _____
Departmental Representative                               Donor

_____
Approval of Director (if required)

1. Department File
2. Donor's Copy
3. Administrative Copy

DMNH 4 (12/77)

Fig. 2.—Donation form designed by the Denver Museum of Natural History using 3-part carbonless paper (original size: 8½ by 11 inches).

situation where one institution sends material to another, until some future date when the second reciprocates.

Exchanges also should be thoroughly documented. An invoice usually accompanies incoming specimens, but if one does not exist, one should be prepared. Exchanges are discussed further in the section on documentation.

## SPECIMEN PREPARATION

Collections acquire both prepared and unprepared specimens. All acquisitions should meet a minimum standard for physical condition and available data. A poorly prepared specimen may be of little use and should not be accepted, whereas an unprepared specimen should generally be prepared before cataloging. Minimum standards of physical condition and data should be described in the institution's collection management manual.

Bird specimens may be prepared in several ways: 1) study skins, 2) museum mounts, 3) fluid-preserved specimens, 4) freeze-dried specimens, and 5) skeletons. Special preparations include frozen tissues and preserved stomach contents. Other preparations that should be considered are eggs and nests. The preparation method chosen will depend upon the intended use of the specimen. A bird which has lost too many feathers and is too badly decomposed for preparation as a study skin may be adequate for preparation as a skeleton.

### Data

Regardless of the method of preparation, professional standards for the amount of data that accompanies specimens are the same. Those data generally are recorded in one of two places, if not both, when the specimen is first collected: 1) field notes, and 2) specimen tags.

All collectors should keep accurate field notes, written in permanent ink on 100 per cent cotton rag bond paper. As described by Hall (1962), Anderson (1965), and Remsen (1977), field notes include a catalog, a journal, and species accounts. The catalog is a listing of the specimens prepared by the collector, including for each specimen such information as date of capture (and date of death if different), collection locality, identification, weight, sex, and reproductive condition. All information that is written on the specimen label should be included in the catalog, as well. The journal describes the itinerary taken by the collector, the habitats visited, collecting methods, species lists, and other pertinent information that may not be included in the catalog and on the specimen tag. Hand sketched maps of collecting sites frequently are included.

Although some collectors record information pertaining directly to an individual species in the journal, many will maintain a separate species account (Hall, 1962; Remsen, 1977). Species accounts include detailed notes on individual species, in specific, such observations as descriptions of vegetation and possible food plants, abundance of species, movements and flight, songs and calls, degree of gregariousness, breeding behavior, foraging behavior, and field descriptions to document sight records of rarities.

In writing field notes, it should be remembered that these become archival documents, describing a particular habitat and its species at a particular time in history. Because it is not possible to predict the future of a locality or its faunal components, the information contained in field notes may prove to be a valuable record for future researchers.

Specimen labels contain most of the data that give the specimen its value. The actual categories of data that should be included vary somewhat with the preferences of individual collectors. However, most collectors will agree that it is better to include too much than not enough, because much of the data can be obtained only when the specimen either is alive or very recently killed. Older specimens contain relatively few categories of data when compared with recently collected material.

Specimen labels should be made of 100 per cent cotton rag, water-resistant paper, such as Byron Weston Resistall Linen Record (Fink *et al.*, 1978). Ink used to write on tags should be black permanent ink that is insoluable in water, formaldehyde, ammonia, and alcohols (Hall, 1962; Anderson, 1965; Fink *et al.*, 1978). Scientific names for study skins generally are written in pencil. The exact arrangement of data on the tag is not crucial, but it should be standardized for each collection to minimize omission of data (Fig. 9).

The institutions surveyed agreed that the following categories of data should be included on the specimen tag if at all possible:

1. Scientific name
2. Collection locality and elevation
3. Collector
4. Collector's number
5. Collection date
6. Preparator
7. Preparator's number
8. Sex
9. Weight
10. Description and measurements of gonad(s)
11. Presence and size of incubation patch
12. Amount of fat
13. Colors of soft parts
14. Body molt
15. Skull ossification

Other categories of data suggested by individual curators include age data, known parentage and mate, presence of parasites, stomach contents, and special numbers such as frozen tissue numbers. Information about the preparation of the specimen, such as whether or not it was washed, should be recorded (King and Bock, 1978), as well as short descriptions of the habitat and cause of death.

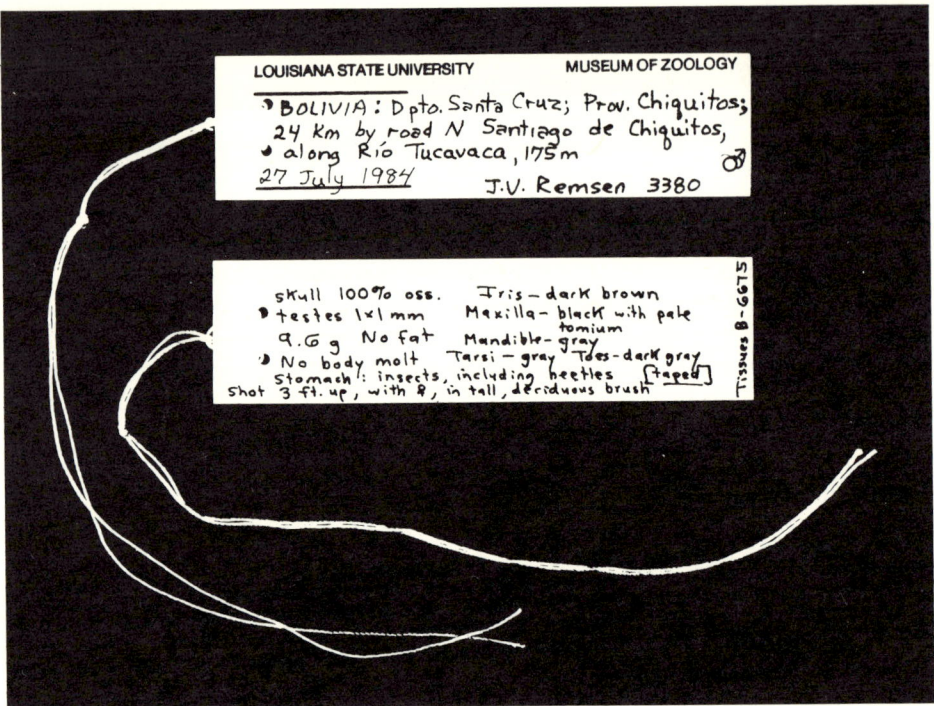

Fig. 3.—Specimen label used at the Museum of Natural Science, Louisiana State University (original size: ¾ by 2¾ inches).

Authorities differ with respect to how one should report localities, especially for remote areas. Riemer (1954) advocated using two direct cardinal compass direction readings, for example, 5 mi. N, 2 mi. W. The methods reported by Axtell (1965) include: 1) odometer readings in road miles or kilometers from a known geographic point to the collection site, 2) airline distance using the nearest cardinal or subcardinal compass directions, and 3) the latitude and longitude of the locality to the nearest minute, second, and fraction of a second. If the latter method is used, collectors also may include the country of occurrence and a reference such as "near Mexico City" to give researchers a general locality (Axtell, 1965). A fourth method is the use of section, township, and range locations. In another, proposed by Hutchinson (1964), the collector measures with a protractor the direction in degrees of the collecting site from a known location, then measures the distance along this line from the known site. Crawford (1983) discussed the use of Universal Transverse Mercator Grids and a decimal geographic grid for North American localities. Sixty-two per cent of the institutions surveyed by the author reported that the format most frequently used for location data was one of the first two reported by Axtell (1965).

Each of these methods has its limitations, and none may be universally applied. Anderson (1965) pointed out that regardless of the method, consistency and accuracy are essential. He noted that the locality should be unambiguous and require little work for curators and researchers to use. A locality designation should use easily cited, easily remembered, and well-known names should be used in the same form by all members of a collecting party. The maps used should be recently printed and accurate. Collectors should make their choice for reporting locality data with these points in mind.

Several of the other data categories included on the specimen label require subjective and qualitative descriptions. Individuals responsible for collecting and directing collection activities may find it advantageous to develop a short set of written guidelines for each data category. Not only would this serve as an aid in training new professionals, but it might help researchers who must interpret the data on the specimen tags. The following descriptions are some of those used at Louisiana State University as prepared by Remsen (1984):

> *Skull ossification.*—"Skull XX% oss." There is little point in using percentages other than the following: 100, 95, 75, 50, 25, 10, 5, 0.
>
> *Gonad data.*—For males, give length and width of testes in mm. For females, record diameter of largest ovum (or "ova minute" for ova less than 1/2 mm dia., or "ovary smooth" if no ova can be distinguished), diameter of oviduct, and length and width of ovary. If you're not 100% sure of the sex, DO NOT GUESS; mis-sexed birds are obviously highly undesirable. Don't be timid about asking more experienced people to sex a bird in question. If the bird cannot be sexed, you can write "gonads rotted", "gonads destroyed by shot", or "no gonads found" to indicate that you didn't forget to look.
>
> *Fat.*—Record amount of subcutaneous fat deposition using the following scale (modified from McCabe, 1943): **no fat**, hardly more than a hint in the dorsal tract or about the pygostyle; **light fat**, a substantial depth, perhaps 1 mm or so in a 20g bird in the dorsal tract; some fat in the furcula; **moderate fat**, quite heavy in the tracts, with small plates elsewhere on the skin, crotch of the furcula fairly well filled; **heavy fat**, considerable amounts of fat removable from many parts of the skin; fat in abdominal cavity; **extremely heavy fat**, deep sheets of fat everywhere between skin and muscle, even over the back; intestines solidly embedded and overlaid, hardly visible.
>
> *Body molt.*—Molting contour feathers, as detected by examination of feather tracts from the inside of the skin, should be noted, with some qualitative statement as to location and extent. If no molt is found, be sure to write "no body molt" (vs. "no molt", which would indicate no wing or tail molt also).
>
> *Soft part colors.*—Give color of iris (eyering and/or facial skin if applicable), bill (maxilla and mandible if colored differently), tarsi, and toes. Use color guide if time and circumstances permit and capitalize color names so used. For series of same species with same soft part colors, reference birds 2-to-x as "soft parts same as JPO 6293." Do not abbreviate colors. Do not record soft part colors for birds that have been dead longer than a couple of hours.

Many researchers insist that months as part of a date should be written out entirely rather than abbreviated to prevent confusion. Some also suggest drawing the pattern of cranial ossification rather than giving percentages or written descriptions.

Study skins, skeletal preparations, and fluid-preserved specimens should all be labeled as completely as possible, with the data described above. Data for eggs and nests should include also the incubation stage and unblown weights of the eggs, the total number of eggs in the clutch, the catalog

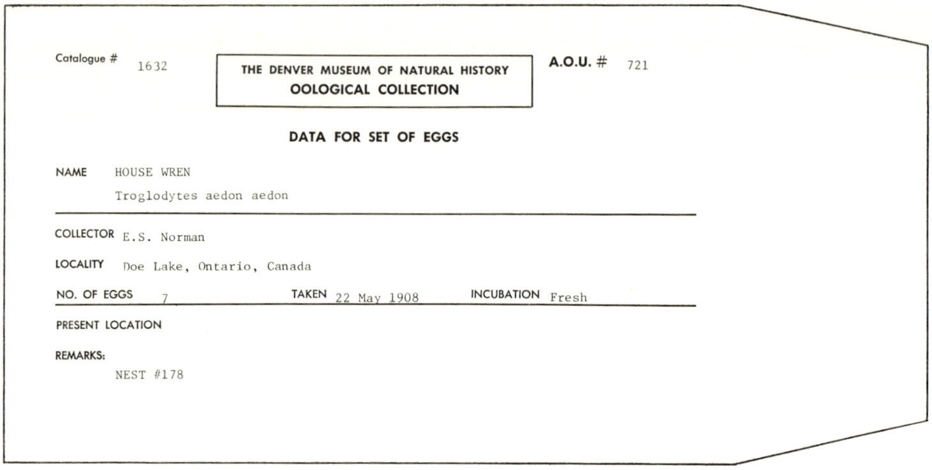

Catalogue # 1632

THE DENVER MUSEUM OF NATURAL HISTORY
OOLOGICAL COLLECTION

A.O.U. # 721

DATA FOR SET OF EGGS

NAME HOUSE WREN
Troglodytes aedon aedon

COLLECTOR E.S. Norman

LOCALITY Doe Lake, Ontario, Canada

NO. OF EGGS 7 TAKEN 22 May 1908 INCUBATION Fresh

PRESENT LOCATION

REMARKS:
NEST #178

Fig. 4.—Envelopes containing egg data are filed by A.O.U. number; Denver Museum of Natural History (original size: 5 by 10 inches including flap).

numbers of the egg's parents or nest builders if collected, and a description of the nest type and materials. The tags stored in the egg or nest containers frequently contain only a minimal amount of information; that is, the scientific name of the eggs or nest builders, catalog number, and locality. To discourage excessive handling of these fragile specimens, additional data should be maintained in separate files (Fig. 4).

### *Study Skins*

Birds prepared as traditional study skins are compact, easy to store, and require little special care beyond routine conservation practices. They retain many of the physical characteristics of the live animal and make comparisons of a series of specimens relatively easy. However, they must be thoroughly clean and well prepared if they are to last for long periods of time.

Several authors have dealt extensively with the preparation of traditional bird study skins (Hall, 1962; Anderson, 1965; Harrison and Cowles, 1970). The instructions for handling and skinning a freshly killed specimen are relatively uniform and will not be included here. The cleaning and stuffing processes, however, are particularly important for the preparation of a long-lasting, useful study skin, and some of the details and variations in those procedures should be noted.

Thorough cleaning of bird skins is essential to their continued usefulness. Grease and dirt from poorly prepared skins can contaminate other specimens either by direct contact in storage trays, or by the alternate handling of greasy and clean specimens. Bloody and fleshy skins are attractive to pests such as clothes moths (Tineidae) and dermestid beetles (Dermestidae). Sixty-four per cent of the institutions surveyed reported the use of both detergents and organic solvents to remove grease, blood, and dirt. Twenty-one per cent used only organic solvents, whereas 14 per cent used neither. Dirty or greasy skins should be washed in cool water with a mild detergent that does not clean by enzyme action. These detergents most commonly include Tide, Ivory liquid or soap, and Joy. Hot water may cause the feathers to fall out, and should therefore be avoided.

After washing, the skin should be rinsed in cool, clean water until all the detergent has been removed, and blotted gently. Care should be taken not to crush the skull or tear the fragile skin. The skin then may be dipped or soaked in an organic solvent such as white gasoline or stoddard's solvent, available from most chemical supply houses. Other organic solvents such as leaded gasoline and Coleman lantern fuel work well, but are not recommended due to their volatile nature. They also contain additional chemicals of which the long term effects on bird skins and feathers are unknown. The length of soaking time will vary depending on the greasiness of the bird. Birds with feathers containing carotenoid pigments (red, orange, yellow, pink) should be soaked only as long as necessary, inasmuch as those pigments dissolve in organic solvents (Schmidt, 1972).

The skin should be blotted once again to remove excessive solvent, then dried with the use of cornmeal or fine, hardwood sawdust. The meal should be worked inside the skin as well as over the outside, and continued until the down is dry and fluffy. The preparator should remove as much of the cornmeal as possible by blowing and gently shaking the specimen; it is frequently necessary to turn the limbs and neck wrong side out to remove completely the cornmeal. Schmidt (1972) noted that this drying method is not always suitable for birds with highly iridescent plumage unless very coarse cornmeal is used, and included alternative methods in his paper.

If the skin of the bird dries out too much during the fluffing step, it should be lightly moistened with either water or a preservative solution prior to beginning the stuffing process. Although arsenate compounds were formerly the main preservative used, only one of the institutions surveyed reported their use today. Eighty per cent of those surveyed reported using nothing at all, whereas 5 per cent use a borax solution.

Stuffing, positioning, and preening should be finished before the skin becomes so dry that feathers are set in undesired positions; thus, it is advisable to have all equipment ready before beginning the stuffing process. A support stick should be used inside the study skin to help prevent broken necks and to provide an anchor to which the legs and specimen tag can be tied. In addition, the tail should be tied to the stick, and the legs and wings tied together inside the skin. In that way, should either the tail or an appendage break off, it would not become totally separated from the skin before repairs can be made.

Both Hall (1962) and Anderson (1965) suggested that study skin bodies should be made by wrapping the support stick with cotton to the approximate original size of the bird's body. For many preparators, however, wrapping cotton in this manner produces a skin with a rounded back that rolls from side to side with any movement of the storage tray. This causes needless wear of a skin, and it can be eliminated by wrapping the stick only to the original size of the bird's neck, then adding more cotton pieces beside and on top of the wrapped stick after it has been inserted into the bird skin (Harrison and Cowles, 1970). This produces a skin with a flat back that will not roll as easily in the tray.

Feathers should be straightened, smoothed and positioned for drying. Proper closure of the beak is important. Tying the beak shut by running a thread through the nostrils is discouraged by some collectors because this may deform the nostril (Van Tyne, 1952).

Some preparators advocate wrapping the finished skin with cotton batting to secure the feathers as they dry (Hall, 1962; Anderson, 1965). This method, however, does not allow for direct scrutiny or alteration of feather positions as drying progresses. Harrison and Cowles (1970) recommend the use of a paper tube to hold the feathers and skin in position while drying. The tube is formed first, and the specimen then slid into it. A third method involves the use of strips of heavy paper and pins. Once the skin has been sewed closed, and the wings positioned, the bird skin is placed on its back on a pinning board. Pins should then be used to hold the beak closed, the wings and toes in place, and the support stick stationary. It is advisable to put strips of heavy paper between the bird's wings and the pins to distribute the pressure of the pin, and to prevent creasing of the feathers. The feathers of the breast and head may then be preened without disturbing the position of the wings. Tails should be pinned open, although not beyond the lateral margins of the body, to permit easy examination of rectrices.

Care should be taken to produce a symmetrical specimen with patterned or colored feathers such as breast spots or wing bars in the correct positions. Harrison and Cowles (1970) discussed methods for handling problem specimens, such as those with large bodies, long necks, long legs, and crests.

Bird skins should be dried in a well-ventilated area free of flies, clothes moths, and other pests. Exposure to direct sunlight should be avoided to prevent rapid bleaching of feather colors.

Specimen tags should be attached to study skins by tying them to one leg for large birds, around both crossed legs and the support stick for small birds. Tags tied to one leg of small birds may cause the leg to be pulled off with frequent handling of the tag.

An alternative method of study skin preparation was presented by Norris (1961), in which the skin and most of the skeleton is salvaged. For this "pelt" or flat mount the bird is skinned, leaving the appendicular bones of one side with the skin whereas all other bones are removed and prepared as a skeletal preparation. The beak may be left with the skin, or removed and left with the skull and skeletal preparation. The flat skin is then glued to a card after being cleaned and degreased as usual. See Norris (1961) for complete instructions.

The advantage of the Norris (1961) method is that many birds can be stored in a small space, and preparation time may be less than for conventional study skins. The skin is subject to less wear because collection users handle the card to which the skin is glued, and not the skin itself (Norris, 1961). However, this method does not allow for undertail coverts and the underside of the tail to be visible unless a window is cut in the card. It is also impractical for large birds. The usefulness for research of a skin of which the bill has been removed with the rest of the skull may also be diminished.

Three additional methods for saving parts of the same bird as skin and skeleton combinations were outlined in Johnson *et al.* (1984). Preparation of bird skins as either traditional study skins or flat skins is primarily based on the individual collector's preferences and research requirements; the majority of the collections surveyed by the author prefer the traditional study skins.

## Mounted Specimens

Mounted bird specimens are most valuable for exhibition and teaching purposes because they are more lifelike and aesthetically appealing than are study skins. Specimens to be used for research purposes are rarely prepared as mounted specimens; mounted specimens require more storage space and are awkward to handle. Specimens prepared as museum mounts are less commonly used for research because they frequently lack collecting data.

The preparation of mounted specimens requires a thorough knowledge of bird anatomy and some artistic ability. Detailed instructions are beyond the purpose and scope of this paper, and readers are referred to any of several taxidermy texts on the market today. Schmidt (1972) presented detailed instructions and special techniques for mounting specimens.

If data are available for specimens prepared as museum mounts, care should be taken to see that they are not lost or separated from the specimen. A catalog number should be written inconspicuously on the base to which the bird is mounted, and all data should be recorded in full in the catalog ledger.

Some collections consider it desirable to relax mounted specimens for which data are available. This must be done very carefully so as not to damage the specimen; in fact, of the collections surveyed by the author, several prefer to leave rare and unusual specimens as mounts rather than take the chance of damaging the specimen. The University of Kansas and the Field Museum of Natural History are among those which have successfully relaxed a number of their mounted specimens.

## Freeze-dried Specimens

During the last 25 years, bird specimens have been prepared successfully by freeze-drying. This is accomplished by freezing the specimen at temperatures below $-10°C$, then lowering the air pressure around the specimen until it is below the vapor pressure of the ice inside the specimen. This procedure causes the ice inside the specimen to sublimate from the tissues (Hower, 1970). Once freeze-drying is complete, it is necessary to degrease the birds. Specialized equipment is necessary but may be readily constructed (Meryman, 1960, 1961; Hower, 1979). Specimens retain their shape, but when necessary, may be positioned with the use of wire supports and may therefore be prepared for exhibition by this method (Hower, 1970).

None of the institutions surveyed by the author reported using this method for study skin preparation. However, the University of Arkansas reported using it during the late 1960's (D. James, personal communication). Constructed from a chest-type deep freezer, a vacuum pump and a vacuum gauge, the chamber was big enough to hold up to 50 small songbirds at one time. Although it took one to two weeks of constant running to prepare those specimens, the set-up could run unattended, cutting the human effort to a fraction. Once the preparation has been completed, normal care for study skins is sufficient for the protection of freeze-dried specimens.

Disadvantages of the freeze-dried method include maintenance and repair of the vacuum pump, and the effect of the degreasing process that follows freeze-drying. The color of the beak, feet and legs of birds may be bleached out by that process.

### Fluid-preserved Specimens

Fluid-preserved bird specimens are most frequently used in anatomical and histological studies. They should be prepared as soon as possible after death in order to preserve the tissues in as natural a state as possible. Use of a fixative such as 10 per cent formalin stops enzymatic action in the tissues and stabilizes the cellular structure of the tissues.

Before fixing a specimen, it should be washed gently with alcohol or a detergent to destroy the water repellent properties of the feathers and ensure penetration of the fixative through the feathers to the skin (Berger, 1955). Ten per cent formalin is the most commonly used fixative (Berger, 1956; Quay, 1974; Williams et al., 1977; Fink et al., 1978). It should be injected directly into the body cavity, and for larger birds, into major muscle masses, toe pads, and tarsal skin. The whole bird is then soaked in the fixative until the pink color of the skin and other tissues is completely gone (Quay, 1974). For larger birds, this may require 6 weeks.

Specimens remaining in formalin for long periods of time will decalcify. To counteract this, the formalin should be buffered with either calcium carbonate, a mixture of salts, or borax (Fink et al., 1978). Berger (1956) and Quay (1974) suggest the use of 4 grams monobasic sodium phosphate monohydrate ($NaH_2PO_4H_2O$) and 6.5 grams dibasic sodium phosphate anhydrate ($Na_2HPO_4$) for each liter of 10 per cent formalin.

Care should be taken to avoid fixing the specimen in a contorted position. If a specimen is to be stored in a plastic bag, claws and the bill should be wrapped in either cotton or cheesecloth to prevent puncturing of the bag.

After fixation is complete, specimens should be washed in water to remove excess fixative, then stored in solutions of either isopropanol or ethanol. Opinions differ with respect to which alcohol is better. Isopropanol is preferred by many users because its initial cost is usually less, it can be diluted more, and it is not regulated by the United States government as is ethanol. Isopropanol is also less volatile than ethanol, reducing the fire hazard and fluid loss (Fink *et al.*, 1978). However, isopropanol is not as water soluble as ethanol and is considered more undesirable to work with because of its smell and the allergic reactions sometimes developed by workers who are exposed to it (Fink *et al.*, 1978). Louisiana State University reported the use of 2 per cent Phenoxetol for a small number of specimens, primarily Recent Louisiana birds and zoo birds, as part of a test to determine the usefulness of this preservative.

Correct shelf storage concentrations of alcohols are still debated as well. Most users agree that concentrations of isopropanol below 45 per cent are dangerous to specimen quality, yet those of 55 per cent or higher may produce brittleness. Collections using ethanol generally use a 70 per cent to 75 per cent solution (Knudsen, 1972; Fink *et al.*, 1978). Ninety-one per cent of the bird collections surveyed by the author use 70 per cent ethanol.

Fluid-preserved specimens should not be crowded into containers. The volume of preservative should be twice the volume of the specimens (Zweifel, 1966). Very large specimens may be stored in stainless steel vats, plastic carboys, and earthenware crocks (Kannemeyer, 1973; Williams *et al.*, 1977).

## *Skeletons*

The preparation of bird skeletons may be difficult due to the fragile nature of most bird skeletons. Care should be taken to ensure that none of the small bones of wings and tails are lost during field preparations and the cleaning process. Methods for preparing and cleaning vertebrate skeletons have been outlined by several authors (Berger, 1955; Hildebrand, 1968; Harrison and Cowles, 1970; Williams *et al.*, 1977).

The first step should be to verify the identification of the specimen. If the identification is in any way doubtful, the plumage should not be removed until the species name can be verified. If possible, the specimen in question should be referenced to a skin of the same species. Unidentified specimens without plumage are useless, and wrongly identified ones may cause additional mistakes and confusion if they are put into a collection (Harrison and Cowles, 1970).

The skeleton tag should be labeled with the same information as a skin tag. Some institutions request that the total length, tail length, tail molt data, wingspan, and wing chord also be included (Van Tyne, 1952). The feathers and skin should be removed quickly, but with care so as neither to damage or lose delicate bones. Berger (1955) and Harrison and Cowles (1970) recommended leaving the terminal two to three primaries in place to avoid disturbing the small wing bones. Major portions of muscles should be removed from the breast, legs, and thighs to speed the drying and cleaning process. Breast muscles should be severed even on small birds, so that when they dry they will not warp the keel. The skin should be removed from the feet of large birds and the toe pads incised to facilitate cleaning (Berger, 1956). The specimen should be sexed, the gonads measured, and the body eviscerated.

The specimen should be bound into a compact bundle with thread to prevent disarticulated parts and limbs from becoming separated from the rest of the skeleton. The specimen should be dried in a well-ventilated area away from insect pests; a wooden frame screen box protects specimens from flies, ants, and other bugs yet allows the specimens to dry. Quick and complete drying of skeletons prevents maceration and inhibits mold formation on the specimen (Sommer and Anderson, 1974). If the skeletons are to be cleaned by dermestids, insecticides should not be applied to the specimens while they are drying.

The use of dermestid beetles (Dermestidae) is the most efficient and effective method for cleaning vertebrate skeletons (Williams *et al.*, 1977). The beetles eat the flesh, scales and feathers, leaving the bones clean. Although dermestid colonies do require a constant temperature and some maintenance, they are usually not difficult to maintain (Hall and Russell, 1933; Tiemeir, 1940; Russell, 1947; John, 1979; and Valcarel and Johnson, 1981). Eighty per cent of the institutions surveyed reported the use of a dermestid colony.

If completely articulated skeletons are desired, care must be taken when they are to be cleaned by dermestid colonies. If the bones of the feet are to remain articulated, a 50 per cent formalin solution should be brushed on and allowed to dry to prevent the dermestids from eating scales and ligaments (Sommer and Anderson, 1974). If specimens are left too long in a colony, the bones will become fully disarticulated. For this reason, some collections prefer to prepare skeletons by maceration.

Damage to tags by dermestids may be prevented by the application of formalin. Specimens should be placed in the colony between two layers of cotton. The cotton provides adult dermestids with an excellent place to lay eggs, and it enhances travel by the beetles between specimens (Sommer and Anderson, 1974). When specimens are removed from the colony care must be taken to retrieve all bones.

Once the skeleton is clean, it should be fumigated thoroughly to kill any dermestids still attached. After fumigation, greasy bones and those with tissue still attached may be soaked in a solution of ammonium hydroxide (3 parts water to one part concentrated ammonium hydroxide) for 6 to 12 hours followed by soaking in clean water for 12 to 24 hours (Williams *et al.*, 1977). An alternative is to soak bones in trichlorethylene until degreased (Sommer and Anderson, 1974). Bones may become dangerously softened if left in ammonium hydroxide and water too long, so caution must be used. Any remaining tissue is removed by scraping or picking with dissecting instruments. The cleaned skeleton should be thoroughly dried to prevent mildew and mold, then placed either in small boxes or capped glass vials. It is essential that specimen tags remain with the skeleton during all phases of preparation.

## Eggs and Nests

Many bird collections contain collections of nests and eggs (Kiff, 1979). Sixty-nine per cent of the institutions surveyed by the author reported egg collections, but of those, only 11 had acquired more than 100 sets of eggs or nests since 1973 (as reported by Banks *et al.*, 1973). Only 5 of those had acquired more than 1,000 sets.

Bird nests require little preparation beyond a thorough fumigation and accurate labeling before being placed in boxes. If a nest is too large to be collected in its entirety, the height and diameter should be measured, and only the central portion collected. Representative material from the remaining portion of the nest also may be collected and retained. Notes should be made on the position, construction and shape of the nest, and a correct identification of the bird that was using the nest is essential. Nests should be labeled in a similar fashion to skins, with the tag tied directly to the nest (Harrison and Cowles, 1970).

Eggs should be prepared by removing the contents through a single hole. This hole should be as small as possible to avoid weakening of the shell and to ensure the egg's usefulness to researchers who wish to measure egg shell weights (Kiff, 1978). Harrison and Cowles (1970) point out, however, that it is better to make a larger hole and ensure that the shell is clean than to make a tiny hole through which the contents will not completely pass.

To begin the hole, a single puncture should be made near the center of the egg at its widest point using a needle or pin. The puncture should be on the side with the fewest natural markings. A small drill bit should be used to open the hole. Care must be taken not to force the bit as this may cause the side of the egg to collapse. When the hole is completed, the ragged edges of the shell membrane should be trimmed by rotating the drill lightly in both directions just under the lip of the hole. This removes any obstruction to the flow of the egg contents which may then be forced out by using air jets from a small blowpipe.

Embryos must be disarticulated before they may be removed from the egg. Small and intermediate-sized embryos may be broken up by first removing as much liquid as possible from the egg, then reintroducing water with the blowpipe and forcibly mixing the water and egg contents. Alternating this process with blowing is usually successful. Large embryos may be removed by treatment with pepsin, a protein-digesting enzyme. An aqueous solution of this enzyme is injected into the embryo. Care should be taken to prevent spilling of the solution onto the egg's surface as it may damage pigmentation. The contents of the egg should be kept moist and well

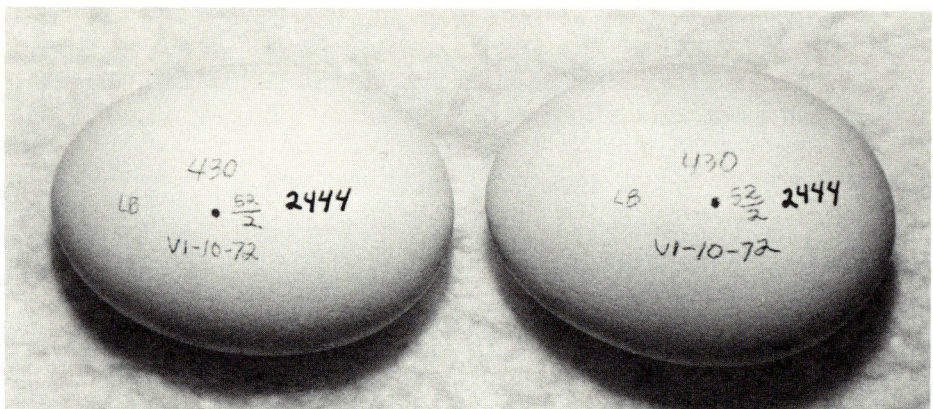

Fig. 5.—Data marked on eggs in standardized positions relative to blowhole (clockwise starting at top, in pencil): A.O.U. number, collector's number/number of eggs in clutch, date collected, and collector's initials. Catalog number is in ink at right.

supplied with enzyme. Every two to three days, part of the egg's contents may be blown out until all the embryo is removed (Kiff, 1978).

The egg shell should then be repeatedly rinsed with water until clean. Any residue left behind may cause the egg shell to deteriorate. The egg should be dried by leaving it hole downward on a blotter.

Once the egg is dried, it should be marked so that it may be identified if separated from its label. The precise form of marking eggs varies from collector to collector, but the following data should be included (Kiff, 1978):

1. Collector's number—each clutch receives a single number
2. Number of eggs in the clutch—often recorded beneath the collector's number with a slash between the two numbers
3. Date of collection—include the month in Roman numerals, day and year
4. AOU number of the parent species
5. Collector's mark or initials

All numbers should be written legibly directly on the egg with permanent black ink. The information is generally written as small as possible in standardized positions near the blowhole (Fig. 5). Harrison and Cowles (1970) and the Smithsonian Institution Leaflet (1957) provide some alternative methods and additional techniques for egg preparations.

### Special Preparations and Associated Materials

### Parasites and Stomach Contents

Complete preparation instructions are outlined by Watson and Amerson (1967), Harrison and Cowles (1970) and Pritchard and Kruse (1982), but in general, endoparasites, ectoparasites, and stomach contents are usually preserved in formalin, isopropanol, or ethanol. Endoparasites such as tape worms (Class Cestoda) may be preserved adequately in a 5 per cent formalin solution (Anderson, 1965), whereas ectoparasites such as feather lice (Order Mallophaga) are best preserved in a 70 per cent ethanol solution (Borror *et al.*, 1981). Stomach contents should be preserved in 70 per cent ethanol as well. All materials should be permanently labeled with the scientific name and collector's number for the bird specimen from which they are taken, as well as an indication of the age and sex of the bird, the collecting locality, and the date of collection (Harrison and Cowles, 1970).

## Sound Recordings

"Recording bird sounds is both a science and a skillful art" (Gulledge, 1977). Care must be taken to capture sounds accurately with minimum distortion and to document the recordings thoroughly in order to ensure their scientific value. Gulledge (1977) discussed equipment needs as well as recording techniques that would be useful to both the scientist and the hobbyist. McWilliams (1979) presented a thorough discussion of tapes and equipment relative to their long term storage capabilities and preservation requirements.

Thorough documentation is essential to the scientific value of a sound recording. Hardy (1984) emphasized the importance of treating the collection of sounds as systematically and in as disciplined a manner as the collection of the more traditional specimens. According to Gulledge (1977) and Simms (1979), the types of information that should accompany each recording may be divided into four categories:

1. Identification and locality
    —species identification
    —field guide reference used to make identification
    —whether identification was made visually and/or by song
    —specific locality
    —date and time of day
2. Behavior and biology
    —observed behavior associated with sounds
    —whether another recording was used as a lure
    —presence of other animals
3. Habitat data
    —description of physical environment
    —weather conditions
4. Technical data
    —name and address of recordist
    —description of equipment used (tape, recorder, microphone system, accessory equipment)
    —playback instructions if applicable
    —account of background sounds

The Library of Natural Sounds at Cornell University has developed a standardized form in order to facilitate documentation in the field of recordings, as well as subsequent cataloging of recordings in a collection

(Gulledge, 1977). These forms plus instructions for their use are available from the Library of Natural Sounds (Gulledge, 1979).

Hardy (1984) suggested recording much of those data directly onto the tape. At the beginning of a recording session, the recordist should dictate the tape number, his name, the locality, date, time, weather, habitat, and type of recording equipment. A short blank should be left before beginning to record sounds.

Each song or unit of sound should form a distinct unit, beginning with the name of the species being recorded (Hardy, 1984). The cut should include a complete uninterrupted sequence of the sound, then end with data relating to the sounds. This alleviates some of the problems involved with editing and cataloging of the cuts, and decreases the chance of separating a recording from its data.

Gulledge (1977) recommends either of two methods of editing in order to make the recordings on each tape more accessible and useful. The first method consists of two steps: 1) make a copy of the original and 2) cut the copy into segments according to species, and splice the segments onto tapes assigned to each species. This method, although time consuming, makes the collection of recordings easier to use. For the second approach, one tabulates the contents of the tape and maintains indices to reference the information. Some institutions, such as the Florida State Museum, use a combination of both methods.

Editing tapes for the purpose of publishing recordings on discs requires some special considerations, which were summarized by Hardy (1978).

## Slides and Photographs

Photographic records of birds should be documented as thoroughly as is possible in order to ensure their scientific value. Data should include species identification, locality, date, behavior, photographer's name, and any technical information that might affect the image (filters, type of film, and the like.

## Frozen Tissues

In 1983, the National Science Foundation sponsored the Workshop on Frozen Tissue Collection Management. The proceedings of this workshop detail some simple methods for collecting tissues under field conditions in addition to other matters of curatorial concern (Dessauer and Hafner, 1984).

## DOCUMENTATION

If the value of a collection is to be realized, it must be documented and organized in such a manner as to allow efficient and complete retrieval of specimens and information. A collection loses value if specimens cannot be found and data are lost. Regardless of the size of the collection, it should be possible to locate any specimen and its data at any time. The easiest way

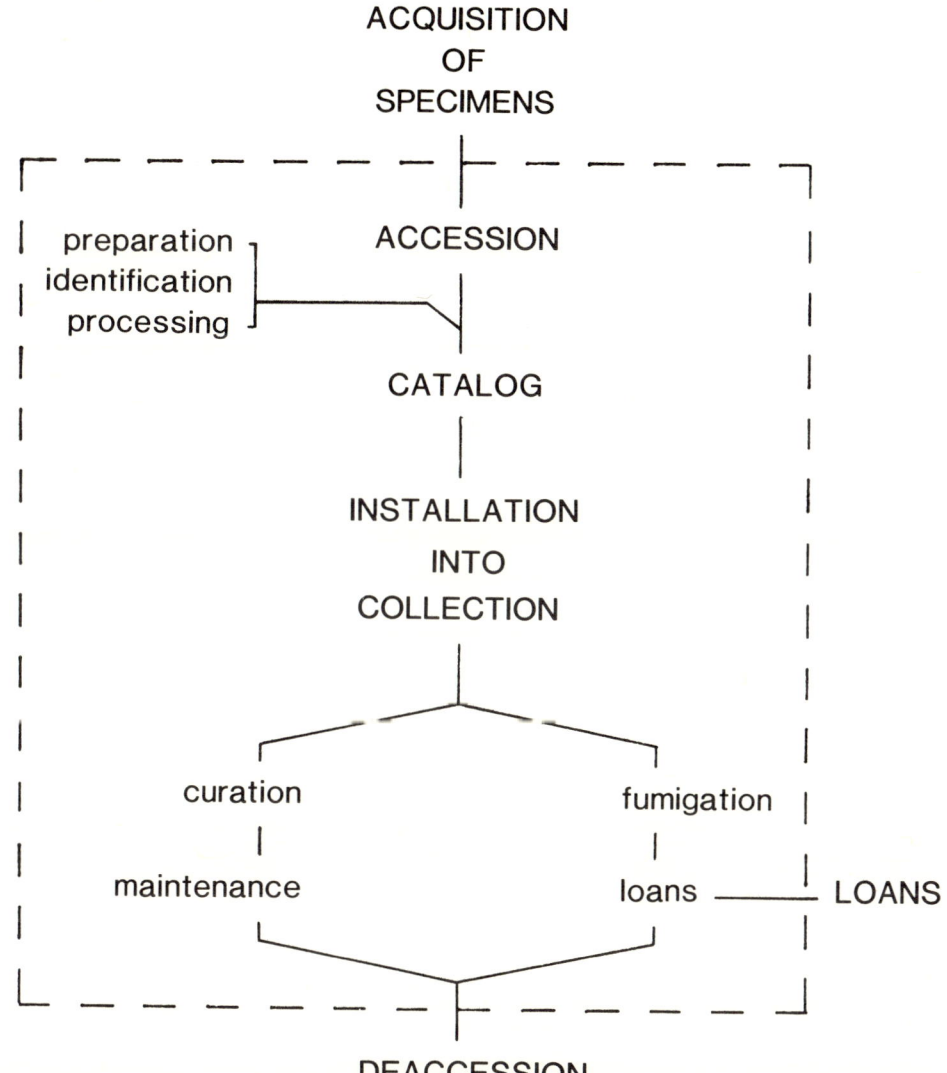

Fig. 6.—Phases of collection management process. Dashed line signifies institutional boundaries.

to keep track of an individual specimen is to assign a number that remains with the specimen until it is destroyed.

Documentation is part of the overall collection management process (Fig. 6) and includes: 1) accessioning, 2) cataloging, 3) loan procedures (discussed in section on collection use), and 4) deaccessioning (Guthe, 1970; Dudley *et al.*, 1979). A standard written procedure for each part of the process eliminates guesswork and decreases the probability of errors.

*Accessioning*

A specimen should be accessioned as soon as it arrives at the institution. Accessioning is the process of assigning a number to each acquisition and serves to document initially the components of the acquisition. An acquisition may consist of either a single specimen or a group of specimens acquired from a single source.

The accession catalog may be maintained either in a ledger, preferably bound, or on cards (Figs. 7-8). Entries should be written with a permanent, waterproof, black ink, and if typed, use a fabric carbon ribbon. The accession numbers may be a simple sequential numbering system or they may be assigned sequentially based on the year of acquisition (for example, 1984-24, 1984-25, 1984-26, and the like). The latter method gives an immediate reference to the year the specimens were acquired, the relative time during the year in which they were acquired, and the number of acquisitions for the year.

The accession catalog should serve as a general reference to what has been acquired, when, and from what source. Names and addresses for donors and collectors should be included, but specific data for each specimen are generally omitted unless the latter are not available on specimen tags, donation forms, and in field notes. Abbreviations should be avoided, particularly for dates and localities, as they may cause confusion in the future. The main categories of information which should be included in the accession catalog are:

1. Accession number
2. Date received
3. Date accessioned or entered into catalog
4. Nature of material
5. Name(s) and address(es) of donor(s)
6. Name(s) of collector(s)
7. How obtained by donor
8. Location and date of collection
9. Catalog numbers for specimens
10. Remarks

Other categories which might be included are entries to indicate the presence of correspondence and field notes, the name of the person entering the acquisition, and the disposition of the specimen if deaccessioned.

The accession number should be entered immediately on the accession and donation form for the acquisition. Specimens which will not be cataloged immediately also should be tagged with the accession number. This step is particularly important for unprepared specimens that are temporarily stored in a freezer. A tag on each specimen decreases the chance of separating the specimen from its data. Accession numbers should be written on tags with permanent, waterproof, black ink.

Following accessioning, prepared specimens should be fumigated thoroughly to prevent infestation of the collection. All specimens should be identified as completely as possible before cataloging.

## Cataloging

Cataloging is the process of assigning a unique number to each specimen. Specimens are referenced by the catalog number in research publications, on loan and exchange forms, and in all cross-reference files. It is the catalog number that makes it possible to locate an individual specimen at any time.

The numerical catalog is a valuable document containing an enormous amount of ·data. As such, it should be protected from casual handling and possible destruction by fires and vandals. One way to protect these data is to store the catalogs in a fire-proof cabinet. Another method is to maintain a back-up copy of the original catalog. Seventeen of the collections surveyed by the author reported the existence of a back-up copy in the form of microfilm (8), cards (4), computer storage (4), a species catalog (1), and photocopy (1).

For any cataloging system, decisions need to be made concerning the numbering system(s) to be used, the physical format for the catalog, and the categories of data to be included in the catalog. These decisions should be based on the needs of the individual collection.

### Numbering System

Of the collections surveyed by the author, 10 used one sequential numbering system to catalog all of the specimens in the collection, regardless of the type of preparation. If a specimen has been prepared for several purposes, such as skin, body skeleton, stomach contents, frozen tissue and the like, all of the preparations would receive the same catalog number, and the specimen label would include a reference to indicate the existence of the other preparations. The numerical catalog also would indicate the various types of preparations.

Seventeen collections reported the use of separate numerical systems for the different types of preparation. However, only four reported a unique numbering system for each type of preparation; the others reported systems that combined certain groups of preparation types. For example, three collections combined skins and skeletal preparations together in one system, with separate systems for fluid-preserved, eggs, and nests. Only three institutions reported the use of prefixes to identify the type of preparation (in one system, "E-" is used for eggs, "N-" for nests, "S-" for skeletons, and the lack of a prefix indicates skins). Those systems appeared to be based on practical decisions and personal preferences for the individual collections.

### Catalog Format

Catalog numbers are sequentially assigned and recorded either in a ledger or on cards (Figs. 9-10). Eighteen of the collections surveyed reported the use of a bound ledger for the numerical catalog, whereas eight used both a bound ledger and a card system. Only one collection reported the sole use of cards. Ledgers should be made using 100 per cent cotton rag paper; cards generally are made with a high quality index card stock. Entries in the catalog should be written with permanent, waterproof, black ink.

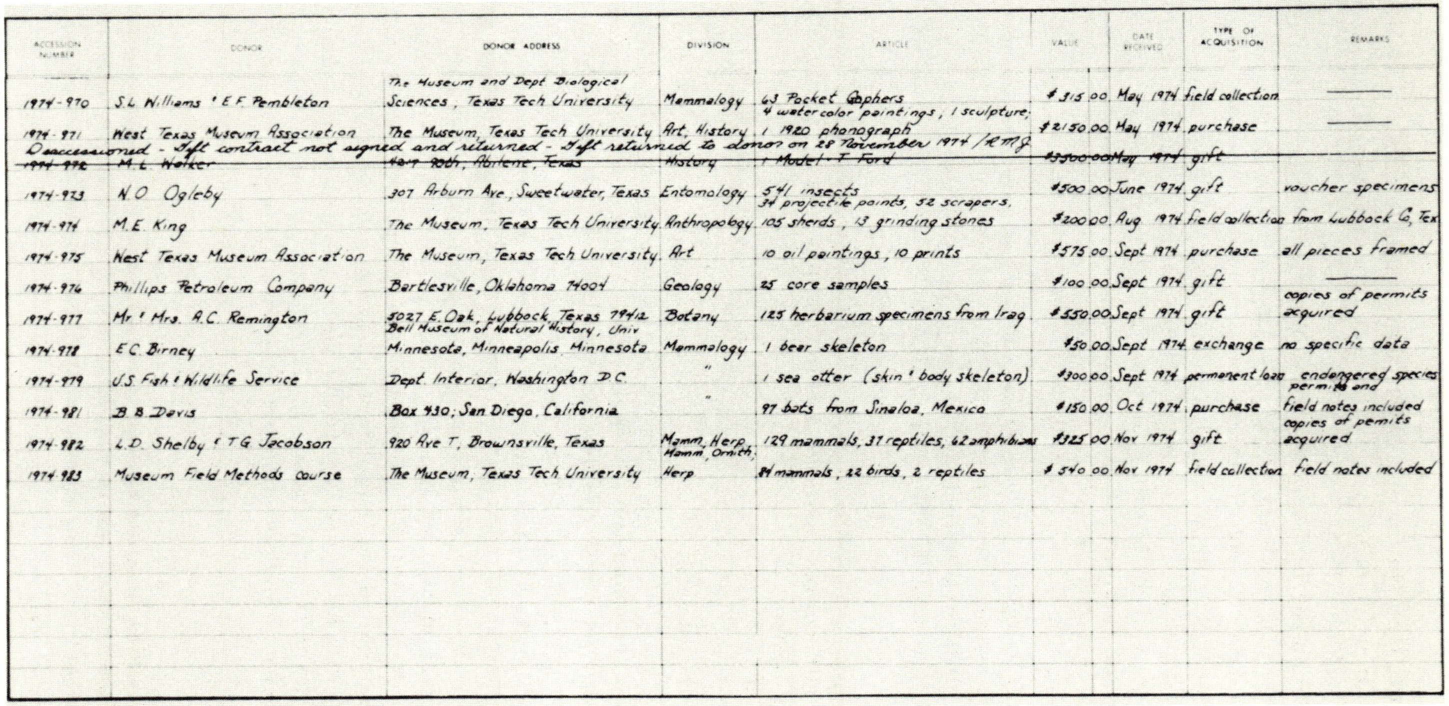

| ACCESSION NUMBER | DONOR | DONOR ADDRESS | DIVISION | ARTICLE | VALUE | DATE RECEIVED | TYPE OF ACQUISITION | REMARKS |
| --- | --- | --- | --- | --- | --- | --- | --- | --- |
| 1974-970 | S.L. Williams & E.F. Pembleton | The Museum and Dept Biological Sciences, Texas Tech University | Mammalogy | 63 Pocket Gophers | $315.00 | May 1974 | field collection | ——— |
| 1974-971 | West Texas Museum Association | The Museum, Texas Tech University | Art, History | 4 watercolor paintings, 1 sculpture; 1 1920 phonograph | $2150.00 | May 1974 | purchase | ——— |
| Deaccessioned - Gift contract not signed and returned - Gift returned to donor on 28 November 1974 / HMJ | | | | | | | | |
| ~~1974-972~~ | ~~M.L. Walker~~ | ~~4019 90th, Abilene, Texas~~ | ~~History~~ | ~~1 Model-T Ford~~ | ~~$3500.00~~ | ~~May 1974~~ | ~~gift~~ | |
| 1974-973 | N.O. Ogleby | 307 Arburn Ave., Sweetwater, Texas | Entomology | 541 insects | $500.00 | June 1974 | gift | voucher specimens |
| 1974-974 | M.E. King | The Museum, Texas Tech University | Anthropology | 34 projectile points, 52 scrapers, 105 sherds, 13 grinding stones | $200.00 | Aug 1974 | field collection | from Lubbock Co, Tex |
| 1974-975 | West Texas Museum Association | The Museum, Texas Tech University | Art | 10 oil paintings, 10 prints | $575.00 | Sept 1974 | purchase | all pieces framed |
| 1974-976 | Phillips Petroleum Company | Bartlesville, Oklahoma 74004 | Geology | 25 core samples | $100.00 | Sept 1974 | gift | ——— |
| 1974-977 | Mr & Mrs. A.C. Remington | 5027 E. Oak, Lubbock, Texas 79412 | Botany | 125 herbarium specimens from Iraq | $550.00 | Sept 1974 | gift | copies of permits acquired |
| 1974-978 | E.C. Birney | Bell Museum of Natural History, Univ Minnesota, Minneapolis, Minnesota | Mammalogy | 1 bear skeleton | $50.00 | Sept 1974 | exchange | no specific data |
| 1974-979 | U.S. Fish & Wildlife Service | Dept. Interior, Washington D.C. | " | 1 sea otter (skin & body skeleton) | $300.00 | Sept 1974 | permanent loan | endangered species permits and |
| 1974-981 | B.B. Davis | Box 430; San Diego, California | " | 97 bats from Sinaloa, Mexico | $150.00 | Oct 1974 | purchase | field notes included |
| 1974-982 | L.D. Shelby & T.G. Jacobson | 920 Ave T, Brownsville, Texas | Mamm, Herp, Ornith; | 129 mammals, 37 reptiles, 62 amphibians | $325.00 | Nov 1974 | gift | copies of permits acquired |
| 1974-983 | Museum Field Methods course | The Museum, Texas Tech University | Herp | 84 mammals, 22 birds, 2 reptiles | $540.00 | Nov 1974 | field collection | field notes included |

Fig. 7.—Format for accession ledger used at The Museum, Texas Tech University (original size: 186 by 395 millimeters).

**Master Accession File**

Texas A&M University
Texas Cooperative Wildlife Collection

| Acc'n No. | Division | Nature of material | Date entered |
|---|---|---|---|
| 1200 | HERPS | DEAN 1982 & 1983 misc. specimens | 27 March 1984 |
| 1201 | HERPS | Dixon etal. April 1982 Red River Co. | 2 April 1984 |
| 1202 | Mammals | 13 Tursiops - salvaged Tx coast | 3 April 1984 |
| 1203 | BIRDS | 1 wing - Surf scoter - USFWS | 26 April 1984 |
| 1204 | Fishes | Exchange - AMNH - sharks | 1 May 1984 |
| 1205 | Herps | Ken King - 1982 - Colorado Co. TX | 2 May 1984 |
| 1206 | Birds | 5 skins - Colorado - H. French | 10 May 1984 |
| 1207 | Herps, Birds & mammals | 26 mammals - alc. / 15 birds    West Texas ; WFS 300 | 1 June 1984 |
| 1208 | Birds | misc. 1984 | 28 June 1984 |
| 1209 | Birds | salvaged by G. Lasley; Hays & Travis Co. | 22 June 1984 |

BIRDS
Accession Catalog

Texas A&M University
Texas Cooperative Wildlife Collection

Acc'n No. 1207 - birds                    Date Rec'd 1 June 1984

Nature of Material 15 study skins - birds

Rec'd From Wildlife & Fisheries 300 class     How Obtained field trip

Address TAMU

Correspondence no                          Field Notes yes

Collector K.A. Arnold                       When Collected May 1984

Locality Texas: Val Verde, Terrell & Presidio Co.

Remarks

Cat. Nos. { Mammals
Birds 11352 - 11366
Rept., Amph
Eggs, Nests }

Date of Entry 15 Aug 1984   Entered by J. Boyd

Fig. 8.—The master accession file documents accessions for all four divisions of the Texas Cooperative Wildlife Collection. An accession card is filled out in detail for each accession (original size: 4 by 6 inches).

**CATALOGUE OF BIRDS, MUSEUM OF ZOOLOGY, LOUISIANA STATE UNIVERSITY**

| CATALOGUE NUMBER | ORIG. NUMBER | NAME OF SPECIES | SEX | SKIN/SKEL | LOCALITY | DATE | FIELD PREP NUMBER | COLLECTOR | REMARKS |
|---|---|---|---|---|---|---|---|---|---|
| 113121 | 3055 | Vireo solitarius | ♀ | skin | LOUISIANA: Cameron Par. Hackberry Ridge, 2 mi. WSW Johnsons Bayou School | 29 October 1983 | 942 | J.V. Remsen | Tissues B-3443 |
| 113122 | 1012 | Vireo olivaceus | ♀ | " | " | 4 September 1983 | " | G H Rosenberg | 3497 |
| 113123 | 3064 | " | ♀ | " | " | 13 November 1983 | " | J V Remsen | 3499 1st specimen for Louisiana |
| 113124 | 3045 | Icterus spurius | ♂ | " | " | 17 November 1983 | " | " | 3485 |
| 113125 | 3071 | Icterus galbula | ♂ | " | Garner Ridge 3 mi W Johnsons Bayou School | 13 November 1983 | " | " | 5297 |
| 113126 | TJD 3201 | Carpodacus purpureus | ♂ | " | East Jetty Woods, 2 mi. S Cameron | 17 December 1983 | " | [coll. by J.V Remsen] Tristan J Davis | 5310 |
| 113127 | 1038 | Vireo griseus | ♀ | " | " | 15 October 1983 | " | G H Rosenberg | 3488 1st Louisiana specimen |
| 113128 | 3092 | Parula pitiayumi | ♀ | " | Holly Beach | 18 December 1983 | " | J.V Remsen | 5302 |
| 113129 | GHR 1056 | Pluvialis squatarola | ♂ | " | 3 mi. W Holly Beach | 29 October 1983 | " | [coll. by AP Capparella] G H Rosenberg | 3480 |
| 113130 | LH 3133 | Calidris maritima | ♂ | " | Plaquemines Par, ½ mi. S Ft Jackson | 12 January 1984 | " | J.V. Remsen | 5327 2nd Louisiana specimen |
| 113131 | LH 1046 | Caprimulgus vociferus | ♀ | " | " | 4 December 1983 | " | [coll. by J.V Remsen] Linda Hale | 1st specimen for Louisiana |
| 113132 | 3086 | Contopus virens | ♀ | " | " | " | " | J.V. Remsen | 5320 |
| 113133 | 3087 | Dendroica fusca | ♀ | " | " | " | " | " | " |
| 113134 | 3088 | Dendroica castanea | ♂ | " | " | " | " | " | |
| 113135 | 3089 | Dendroica virens | ♂ | " | Venice | " | " | " | |
| 113136 | 3131 | Zonotrichia (Melospiza) lincolnii | ♂ | skel | Cameron Par, Hackberry Ridge 3mi. WSW Johnsons Bayou School | 26 November 1983 | " | " | |
| 113137 | — | Stellula calliope | sex? feathers only | | St John the Baptist Par, Reserve | | " | Nancy L. Newfield | 2nd Louisiana specimen |
| 113138 | DW 1395 | Sayornis phoebe | ♀ | skin | TEXAS: Schleicher Co., 25 mi. E Eldorado on US Highway 190 | 4 June 1983 | 955 | [coll. by C.C. Wiedenfeld] D. Wiedenfeld | |
| 113139 | 1394 | Zonotrichia leucophrys | ♀ | " | Lubbock Co., 2 mi E FM East 400 on Farris Road, ca 10mi E Lubbock | 23 October 1983 | " | D. Wiedenfeld | |
| 113140 | 1393 | Chordeiles acutipennis | ♀ | skel | Tom Green Co., San Angelo, intersection of Pecan End and Christoval roads | 24 May 1983 | " | " | |

Fig. 9.—Page from catalog ledger used at the Museum of Natural Science, Louisiana State University (original size: 12½ by 17 inches).

| | | Department Catalogue | | | | Texas A&M University |
|---|---|---|---|---|---|---|

Acc'n. No. *1206; 1207*   Date of Entry *18 June 1984* Entered by *J. Boyd*

General Locality *Colorado ; Texas*

| Dept. No. | Orig. No. | Name | Date | Collector | Exact Locality |
|---|---|---|---|---|---|
| 11349 | NRF - AK 2662 | ♂ *Leucosticte arctoa australis* | 21 Jan 1984 | N.R.French | Colo; Larimer Co.; Livermore, Roberts Ranch |
| 11350 | NRF - AK 2663 | ♂ " " " | " | " | " |
| 11351 | NRF - AK 2664 | ♂ " " " | " | " | " |
| accn 1207 11352 | KAA 5545 | ♂ *Hirundo pyrrhonota tachina* | 16 May 1984 | K.A. Arnold et al | Tx: Terrell Co; 10 mi SE Sanderson (Hwy 90) |
| 11353 | KAA 5541 | ♂ *Carduelis psaltria psaltria* | 15 May 1984 | " | Tx: Val Verde Co; ca. 5 km S. Del Rio |
| 11354 | KAA 5540 | ♂ *Geothlypis trichas* | " | " | " |
| 11355 | KAA 5542 | ♀ " " | " | " | " |
| 11356 | KAA 5539 | ♀ *Oporornis (=Geothlypis) tolmiei* | " | " | " |
| 11357 | KAA 5538 | ♂ *Arremenops r. rufivirgatus* | " | " | " |
| 11358 | ? 003 | ♂ *Chondestes grammacus strigatus* | " | " | " |

Fig. 10.—Catalog card maintained by the bird division, Texas Cooperative Wildlife Collection (original size: 4 by 6 inches).

## Categories of Data

Data entries made for each specimen form the basis of future information retrieval; therefore, the categories chosen should reflect the collection's projected uses. A small number of entries per specimen speeds up the cataloging process and the inclusion of more categories might seem redundant inasmuch as the data are available in field notes and on specimen labels. However, this philosophy results in increased handling of the specimens and their labels, and therefore, increases their rate of deterioration. A larger number of catalog entries per specimen allows researchers to determine exactly which specimens meet the criteria for their research project without actually handling every specimen. It also is easier to generate cross-reference files, both manual and computer, from a complete catalog than from the specimen labels.

Catalog entries and the data for each should be standardized. This increases the accuracy of recorded information and facilitates the future transfer of data to a computerized system, even if one presently is not planned. The following categories of information are included in numerical catalogs for study skins for more than half of the collections surveyed by the author:

1. Catalog number
2. Genus
3. Species

4. Subspecies
5. Sex
6. Country
7. State
8. County
9. Specific locality
10. Collection date
11. Type of preparation
12. Collector
13. Collector's number
14. Preparator
15. Preparator's number
16. Accession number
17. Remarks (including associated specimens, special numbers, references to state records, and the like)

Other categories might include family, measurements, age, habitat, capture method, data cataloged, and donor.

Examples of how data should be standardized include entries for dates and types of preparation. Most institutions prefer to have dates written out, not abbreviated; numbers used to represent months can be too easily misinterpreted. Preparation types also may be written out, but a simple set of abbreviations might be developed, such as those used at the Denver Museum of Natural History (based on Williams *et al.*, 1977):

AL—alcoholic
SO—skin only
SB—skin with skull and partial skeleton
KB—skin with skull and complete skeleton
SN—skeleton only
BS—body skeleton (excludes skull)
PS—partial skeleton
BM—body mount
OT—other

Catalog entries for skeletal materials, fluid-preserved specimens, eggs, and nests are similar to those for study skins. However, subspecies, sex, and type of preparation generally are omitted for eggs and nests. Reference to additional information, such as nest record cards and egg data envelopes may be made in the remarks column.

Sound recordings may be cataloged by assigning a number to each more-or-less complete tape. At the Florida State Museum, this "master tape" number is assigned consecutively as each new tape is received. A running tabulation of the contents is recorded and organized in loose leaf binders according to the master tape number. A species cross reference is maintained on 3 × 5-inch cards (Webber, personal communication).

## Cataloging Process

In order to improve the accuracy of data transcription during the cataloging process, as well as facilitate information and specimen retrieval, a collection will generally catalog all specimens of a single field trip and acquisition at one time. Prior to assigning catalog numbers, the specimens may be arranged in the same standardized order that is used to organize the collection. As a result, specimens with repetitive data are cataloged in sequence, and specimens of the same species from a particular field trip will remain together in the collection.

The organizational system for specimen storage and cataloging varies depending on the curator's preferences and the strengths of the collection. Most collections arrange their specimens phylogenetically to the level of genus (Barlow *et al.*, 1977). At the levels of species and subspecies, some collections alphabetize their specimens, whereas others rely on a phylogenetic arrangement.

The most commonly used phylogenetic systems are Peters (1934-1979) and Morony *et al.* (1975). Only two of the collections surveyed by the author did not use one or the other at some level. Four reported using Peters (1934-1979) exclusively, whereas six relied solely on Morony *et al.* (1975). Eleven institutions use Peters (1934-1979) in combination with another system, 10 use Morony *et al.* (1975) in a similar manner. During the last seven years, 12 collections have rearranged their phylogenetic order; of these, nine now use Morony *et al.* (1975) and three use Peters (1934-1979). Other systems used in conjunction either with Peters (1934-1979) or Morony *et al.* (1975) are those of Wetmore (1960), Storer (1971), Edwards (1974), and American Ornithologists' Union (1983). One curator suggested the use of the inventory list from Wood *et al.* (1982a, 1982b) because it incorporates material from volumes of the Peter's Check-list (1934-1979) that has appeared after Morony *et al.* (1975).

An example of a standardized organizational system is the one currently used at the Denver Museum of Natural History. Based on a system developed for mammals at The Museum, Texas Tech University (Williams *et al.*, 1977), specimens are arranged:

1. Phylogenetically to subspecies according to American Ornithologists' Union (1957) for North American birds and Edwards (1974) for all others
2. Within a subspecies, alphabetically by country
3. Within a country, alphabetically by state, province or department
4. Within a state, alphabetically by county or parish
5. Within a county, alphabetically by reference point
6. With regard to a reference point, by latitude north to south and within a latitude by longitude west to east
7. With reference to a specific location, chronologically
8. With reference to a specific date, alphabetically by collector

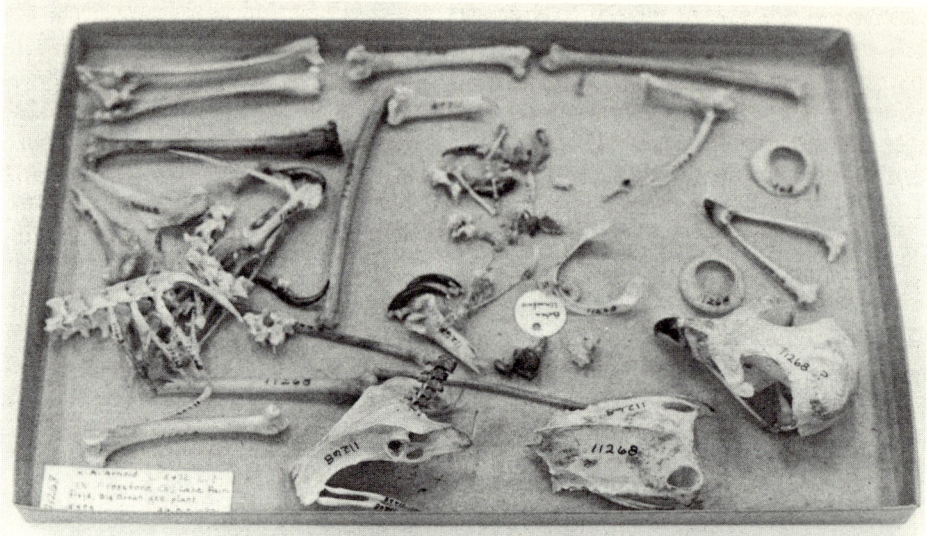

Fig. 11.—The catalog number is 11268; "?" indicates that the sex is unknown.

9. With reference to a specific collector, alphabetically by preparator
10. With reference to a specific preparator, numerically by the preparator's number

Once entry into the numerical catalog is complete, the catalog number should be written with permanent, waterproof, black ink on both sides of the specimen label. For exchange material and specimens with labels other than the institution's own, a blank label with the new catalog number is tied to the specimen. The institution's acronym and the new catalog number are written also on the old label (for example, DMNH 35261, TCWC 2332). Existing labels never should be removed from the specimen. The catalog number also should be written on each skeletal piece (Fig. 11), box and vial labels, labels for wet specimens (Fig. 12), eggs, and egg and nest labels.

When the entire acquisition has been cataloged, the catalog numbers should be entered on the accession card and ledger. Cross reference files may be generated based on taxonomic or geographic units, as discussed beyond in the section on information retrieval.

## Type Specimens

Specimens which have been designated in the literature as either holotypes or paratypes should be specially marked so as not to be confused with other specimens of the same species. These specimens are of particular research value and are generally treated with extra care. The specimen tags for types are frequently color-coded: red for holotypes, and green for paratypes. Some collections also give special designations for state records.

| SKELETON ONLY |
| Texas Cooperative Wildlife Collection |
| 11269    KAA 5499    9May1983 |
| TX:Colorado Co;AttwaterPr.Ch. |
| Refuge |
| Tympanuchus cupido attwateri |
| Wildlife Science, Texas A&M University |

**Texas Cooperative Wildlife Collection**
**Texas A&M University**

Family: _____ ICTERIDAE _____

Species: _____ Molotrus aeneus _____

| Cat. No. | Locality |
|----------|----------|
| 8868 | Mexico:   San Luis Potosi; 33 km W of Valles |
| 8869 | "          "    "     "          "   "  "   "   " |
| 11311 | "     :  Tamaulipas; San Carlos Mtns. |
| | |
| | |
| | |
| | |
| | |
| | |

Fig. 12—Top: box label for skeletal material (original size: 1⅛ by 2⅛ inches). Bottom: label for jar of fluid-preserved specimens (original size: 4 by 6 inches).

## Deaccessioning

Collections generally do not keep all of their specimens forever. Some specimens are discarded due to their extremely poor physical condition; some are exchanged to other institutions for research or exhibit purposes. Space and financial restrictions also may force the transfer of specimens to another institution.

Deaccessioning documents the transfer of all specimens that permanently leave the collection. A written deaccession policy sets guidelines for what can be discarded and how, ensuring compliance with legal and ethical restraints. It also may serve to reassure interested public parties that the institution is not randomly discarding specimens (American Association of Museums, 1978).

Specimens should be deaccessioned only with the written approval of the curator. If the quantity or monetary value of the specimens is great, deaccessioning should be subject to review and approval by a collections committee or the institution's director. This step helps prevent abuses of deaccessioning.

Staff members should not acquire deaccessioned materials, nor can specimens usually be returned to donors. Federal and state laws rarely permit this. Material on permanent loan from the U.S. Fish and Wildlife Service cannot be deaccessioned without their written approval. The legality of a sale of deaccessioned zoological material must be checked carefully with wildlife enforcement officials. As a general rule, sales are strongly discouraged.

Certain specimens, such as types and endangered species, rarely are deaccessioned unless their physical safety is in question. Potential damage resulting from poor collection conditions, such as insect pests, extreme temperatures and humidity, and lack of proper management, may on occasion best be avoided by a transfer of the specimens to another institution.

Exchanges are the preferred method for deaccessions; the specimens may continue to serve a useful purpose at another institution. Specimens without data may add greatly to a teaching collection elsewhere. Frequently, parts of specimens may still be useful for teaching when the entire specimen is not. However, if the physical condition of a specimen is poor enough, the specimen should be destroyed completely.

All deaccessions should be documented fully and supported with no less than the following information:

1. Date of transaction
2. Type of transaction (sale, exchange, and the like)
3. Description of the specimen(s), including catalog numbers and physical conditions
4. Market value of the specimen(s)
5. Reason for deaccession
6. Where the specimen(s) has(have) been relocated

Some institutions use either a deaccession or disposal form to record this information (Fig. 13). This is more common among institutions which have a single institution-wide accession and deaccession system. Of the collections surveyed by the author, only five had deaccession forms. However, all of the collections reported the use of a form for documenting exchanges; 11 used a multipurpose invoice form, and 17 had a specific exchange form (Fig. 14).

A permanent file should be maintained for both deaccession forms and exchange forms. Deaccessions also should be noted in the numerical catalog. A line may be drawn through the record of the deaccessioned specimen, with the date and new location of the specimen noted.

## Storage and Maintenance

The permanency of a collection, as well as efficient specimen use, depends on proper storage and conservation practices. Specimens should be stored under a system that is both logical and practical, considering the uses of the collection and the space and financial restrictions. Sufficient work space should be included in the collection area. In addition, conservation practices should be followed to ensure the survival of the specimens and their documentation.

## Conservation

Ideally, conservation of specimens begins in the planning stage for construction of storage facilities (Cameron, 1968) and exhibits (Hunter, 1974). To prolong the life of a specimen, it is necessary to protect it as fully as possible from destructive agents such as extremes of temperatures and humidity, light, dust, chemical air pollutants, mold, insect and rodent pests, and fire.

The suggested optimum for temperature is $65 \pm 5°F$ (Stolow, 1966a). Temperatures above 70°F permit molds to grow and provide a more attractive environment for harmful insects. Most importantly, a relatively constant temperature should be maintained, making it easier to stabilize the relative humidity and lessening the stress placed on material caused by repeated expansion and contraction in response to temperature extremes. Daily fluctuations should not exceed 15°F (Van Gelder, 1965).

Extreme fluctuations in relative humidity also should be avoided; the optimum range is 50 to 60 per cent (Buck, 1964). Too little moisture in the air causes specimens to dry out and become brittle; too much permits molds to grow. The deteriorating effects of humidity and temperature extremes, as well as preventive practices, have been reported extensively (Kennedy, 1960; Plenderleith and Philippot, 1960; Stolow, 1966a, 1977).

Chemical changes caused by the radiant energy of light may seriously damage specimens (Plenderleith and Philippot, 1960; Feller, 1964, 1968). Both the wavelength and the amount of light cause damage. Infrared radiation can increase significantly the temperature of a small enclosure and cause specimens to dry out (Hunter, 1974). Ultraviolet light causes the greatest photochemical effect, and a rise in temperature of 10°F can double the rate of its activity (Stolow, 1966b). The most obvious result of light-caused damage is the fading of colors.

Enclosed cabinets and the use of filters over fluorescent lights significantly decrease the damaging effects of light. Hunter (1974) and Hubner (1981) discussed the types of filters available, as well as supplier sources for purchases.

Dust and chemical air pollutants can cause both physical and chemical damage. Dust particles act as an abrasive, and repeated cleaning merely adds to the wear and tear of the specimens. Air pollution includes such corrosive agents as sulfuric acid. The best protection against those agents are well-sealed cabinets and an efficient air filtering system for the entire collection area.

Insect pests are an additional problem, making fumigation a necessity. The most common pests are dermestid beetles, obiid (cigarette) beetles,

**DENVER MUSEUM OF NATURAL HISTORY**
CITY PARK
DENVER, COLORADO 80205

**RECORD OF DISPOSITION**                    Nº    222

| DATE | DEPARTMENT | |
|---|---|---|
| TYPE OF MATERIAL | TYPE OF TRANSACTION | MARKET VALUE |
| ☐ accessioned | ☐ exchange | |
| ☐ deaccessioned | ☐ discard | |
| | ☐ sale | |

The following material has been removed from the collections of the Denver Museum of Natural History in ac-
cordance with the policies and procedures set forth in the **Collections Policy Manual.**

REFERENCE #          DESCRIPTION OF MATERIAL (state condition)

WHERE RELOCATED

JUSTIFICATION OF DISPOSITION

_____          _____
Signature of Authorization                 Approval of Director (if value exceeds $500)

DMNH 10 (2/78)          White - Department File  —  Blue - Archives  —  Yellow - Director's Office

Fig. 13.—Record of Disposition form used at the Denver Museum of Natural History (3-part
NCR paper; original size: 8½ by 11 inches).

tenebrionid (confused flour) beetles, carpet and clothes moths, cockroaches,
silverfish, and termites. Some fumigants such as napthalene are merely
repellents and will not kill the pests. Stronger pesticides exist, but many are
not legally available to museums. Many are highly volatile and most are

**DENVER MUSEUM OF NATURAL HISTORY**
CITY PARK
DENVER, COLORADO  80205

**EXCHANGE AGREEMENT**

The Denver Museum of Natural History shall hold unrestricted legal title to any property received pursuant to this Exchange Agreement. All property so received may be utilized in any manner that is in the best interests of the Denver Museum of Natural History.

Department _____ Date_____

Description of material **EXCHANGED:**                                 VALUE: _____

Description of material **RECEIVED:**                                   VALUE: _____

Name and address of recipient: _____

The recipient hereby certifies that the recipient has unrestricted legal title to the property conveyed to the Denver Museum of Natural History.

Remarks:

_____            _____
For Denver Museum of Natural History                          For Recipient

DMNH 11 (2/78)

Fig. 14.—Exchange form used at the Denver Museum of Natural History (3-part NCR paper; original size: 8½ by 11 inches).

toxic to humans to some degree. A recent study (Edwards *et al.*, 1981) discussed the regulations governing the use of pesticides as well as their toxicity. Of the available fumigants, Vapona appears to be the most effective of the least toxic pesticides.

Of the collections surveyed by the author, 18 reported that a fumigant was kept in the cabinets at all times. The most common fumigants for this purpose were paradichlorobenzene (PDB) (10), Vapona (3), and napthalene (3). Eleven collections used a stronger fumigant (Dowfume, carbon disulfide, chlorosol, ethylene oxide) once or twice a year either as a sole fumigant or in addition to PDB, Vapona and napthalene. Edwards *et al.* (1981) should be consulted before choosing a fumigant.

Well-sealed metal cabinets are essential if fumigation is to be effective, and fumigation should be scheduled regularly—two to four times each year. Care must be taken to follow all safety procedures during fumigation, and cabinets should be marked in some way when fresh fumigant is added.

A separate cabinet may serve as a fumigation chamber for problem specimens, incoming loan specimens, and returned loan specimens. Newly acquired specimens must be fumigated immediately after accessioning. One infested specimen can cause the destruction of an entire cabinet of specimens if the fumigation procedures are not rigidly followed.

In addition to general conservation practices, consideration needs to be given to the protection of collections during energy emergencies and natural disasters (American Association of Museums Energy Workshop, 1977; Hunter, 1979, 1980). Advance planning is a necessity to limit the damage as much as possible. There should be readily available a prioritized list of steps to follow and a listing of the individuals responsible for carrying them out. Staff members should know their responsibilities in advance of any emergencies. For example, if the electricity goes off during a severe storm, how long will specimens remain frozen in a freezer before they must be moved, where can they be moved to, and who is responsible for moving them? There is no substitute for written procedures established before an emergency occurs.

For further reading on conservation practices, consult the bibliography by Rath and O'Connell (1977).

## *Organization*

In addition to the collection itself, a collection facility should ideally include work space, library and archive space, equipment and supply storage, and laboratory space for specimen preparations. Facilities should be constructed and organized in such a way as to approach the optimal environmental conditions necessary to preserve specimens.

The research collection usually is subdivided and stored according to the type of preparation. Twenty-six of the collections surveyed by the author reported that each preparation type was stored separately: study skins, skeletal preparations, fluid-preserved specimens, eggs, nests, and the like. Study skins, skeletal preparations, eggs, and nests are most commonly stored in well-sealed metal cabinets (Fig. 15). Most collections reported the use of either open or closed shelving for fluid-preserved specimens. Closed shelving protects specimens from direct lighting.

Fig. 15.—Storage of study skins in metal cabinets, Texas Cooperative Wildlife Collection.

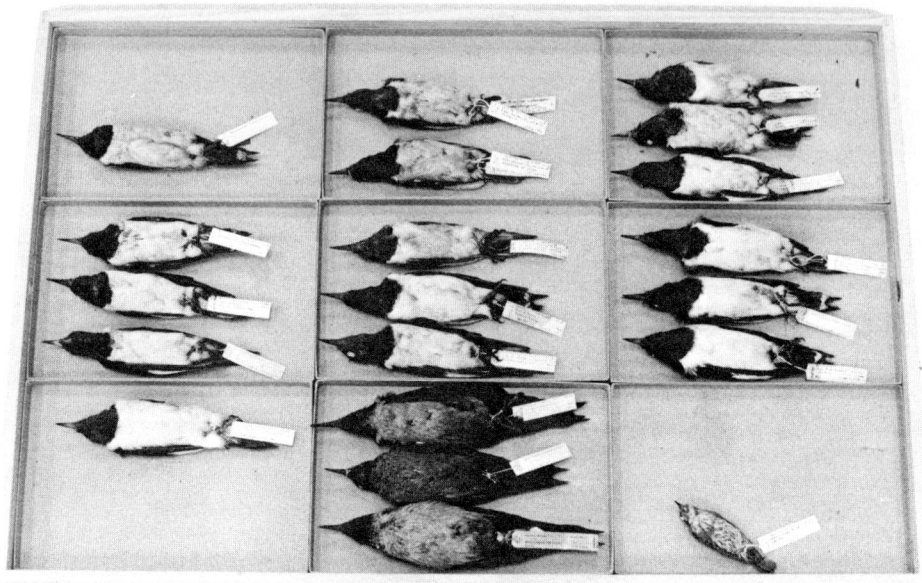

Fig. 16.—Compartmentalized storage of bird specimens. Front of the drawer is to the right.

Several of the divisions at the Academy of Natural Sciences of Philadelphia use compactors to store their collections. Though expensive, compactors are extremely efficient space-savers (Association of Systematics Collections, 1976). For a rapidly growing collection, they may prove to be cost efficient as well. The cost of compactor installation for the Academy of Natural Sciences ichthyology collection in existing space "was less than half the expense of building sufficient additional space, and usable collection space was increased by some 70%" (Fink *et al.*, 1978).

Two possible problems exist with compactors. First of all, standard metal cabinets are too heavy to mount on this type of system, and it is necessary to design a lighter weight cabinet. The second problem is ensuring the airtightness of the cabinet to allow for effective fumigation (Robbins, personal communication).

Most collections reported using the same organizational system for all of the subdivisions of the research collection. As discussed in the section on cataloging, the system should be standardized. A decision should be made concerning specimens with incomplete identifications or data. Should these be placed either at the beginning or end of the appropriate taxonomic level? The arrangement of specimens within a drawer and on a shelf should also be standardized to facilitate installation and retrieval of specimens. For example, the standard order for specimens within a drawer might be from front to back, and left to right.

Type specimens should be stored separately in a locked cabinet. Extinct species frequently are treated this way as well. Also, separate cabinets for

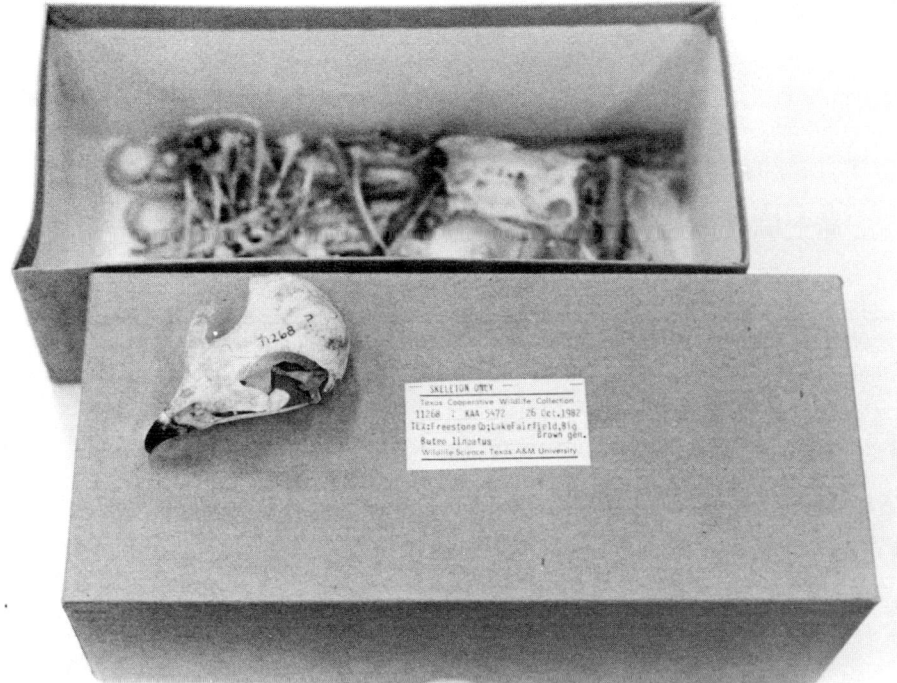

Fig. 17.—Skeletal material is stored in boxes (Texas Cooperative Wildlife Collection).

loaned specimens and specimens used in current research studies should be maintained.

## Study Skins

Drawers should not be so full that specimens damage each other; space also must be left between drawers. Bills and feathers can be broken if allowed to rub against the sides and bottoms of drawers. Small birds will move around less in the drawer if compartmentalized in shallow cardboard trays (Fig. 16). Ideally, these should be made from acid-free materials. Dowler and Genoways (1976) and Kohl (1982) listed suppliers of trays, boxes, vials, and other products useful for collection management.

## Skeletal Material

Skeletal material should be stored either in labeled glass vials or boxes with lids (Fig. 17). Plastic vials crack too easily and frequently are affected by fumigants. Although more expensive initially, glass vials have a longer life span. Vials may be sealed by either plastic or cork stoppers, or cotton. Plastic stoppers often prove to be the most satisfactory inasmuch as cork frequently dries out and becomes brittle, and cotton is easily lost.

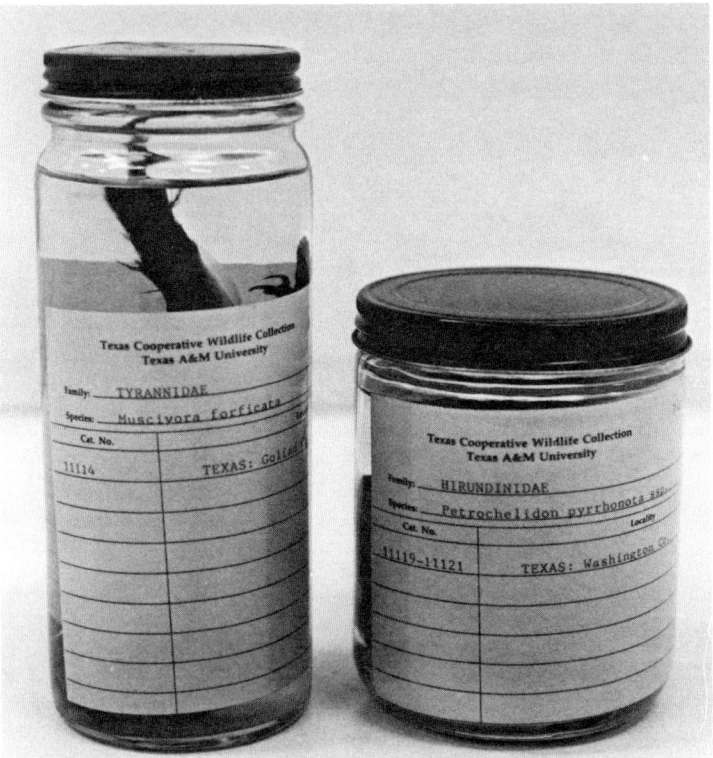

Fig. 18.—Storage of fluid-preserved specimens (Texas Cooperative Wildlife Collection).

## Fluid-preserved Specimens

Specimens preserved in alcohol and other fluids should be stored in glass jars with either screw-top or gasket-type lids. Each jar must be labeled and lids sealed tightly in order to minimize evaporation (Fig. 18). A survey of collection practices in ichthyology collections revealed that lids of polyethylene seem to provide the most efficient seals (Fink *et al.*, 1978). Metal lids rust and do not seal well. Palmer (1974) suggested the use of polyethylene disks in glass canning jars to improve the seal. Styles and brands of both jars and lids are discussed in detail in the "Report on Current Supplies and Practices Used in Curation of Ichthyological Collections" (Fink *et al.*, 1978).

Alcohols are flammable and should be protected from excessive heat. Fluid-preserved specimens also are particularly sensitive to sunlight. If not stored in closed cabinets, shelving should be placed in a location where there is no direct sunlight, and artificial lighting should be routinely left off.

Although evaporation can be slowed with tight-fitting lids, it is rarely eliminated completely. The fluid volume should be checked at regular intervals and topped off when necessary.

## Eggs and Nests

Eggs and nests are very fragile and should be handled as little as possible. Each nest and clutch of eggs should be stored in a box in a cabinet drawer (Figs. 19-20), and removed from the boxes only for measuring and actual comparisons. Egg boxes should be lined with cotton so that specimens do not repeatedly knock each other and the wall of the box. For both eggs and nests, the box should be labeled with the catalog number and species name for easy reference.

## Teaching Collection

If practical, all specimens of a teaching collection should be stored together, regardless of their method of preparation. When choosing an organization for this collection, consideration should be given to the individuals who will be using it most often. Most will not be researchers and therefore not familiar with most phylogenetic systems. A simple and easily understood system might be based on a taxonomic system to the level of genus, with species and subspecies arranged alphabetically. Specimens within a subspecies may be ordered by catalog number. Common names should be listed on cabinet and drawer labels.

Specimens in the teaching collection should be stored and handled with the same care and respect given to research specimens in order to prolong their life span.

## Mounted Specimens

Mounted specimens create unique problems due to the variety of shapes and sizes. A sturdy base is essential to decrease the fragility of the specimen. Most collections surveyed reported using a variety of storage facilities for mounted specimens (cases as well as open and closed shelves). Steel (1970) suggested that passerines and other small to medium-sized birds be stored in a manner similar to paintings in an art collection. With two or three strong hooks, the mounted birds are attached by their bases to a screen on a sliding panel. A series of these panels can be built inside a cabinet, protecting the specimens from both light and dust (Fig. 21).

In lieu of such a system, both small and large birds are stored either in cabinets or on sturdy wooden or steel shelving, away from the light. A drop cloth helps to protect the specimens stored on shelving. For small collections of mounted specimens, it is possible to enclose each specimen individually in plastic. Fumigation, however, is difficult for either method unless the specimens are stored in a room away from work areas.

## Sound Recordings

Sound recordings should be stored in reel boxes in a relatively stable environment (Gulledge, 1977; McWilliams, 1979). Changes in temperature cause expansion and contraction of the tapes, and changes in humidity

Fig. 19.—Storage of nests (Denver Museum of Natural History).

Fig. 20.—Storage of eggs (Denver Museum of Natural History).

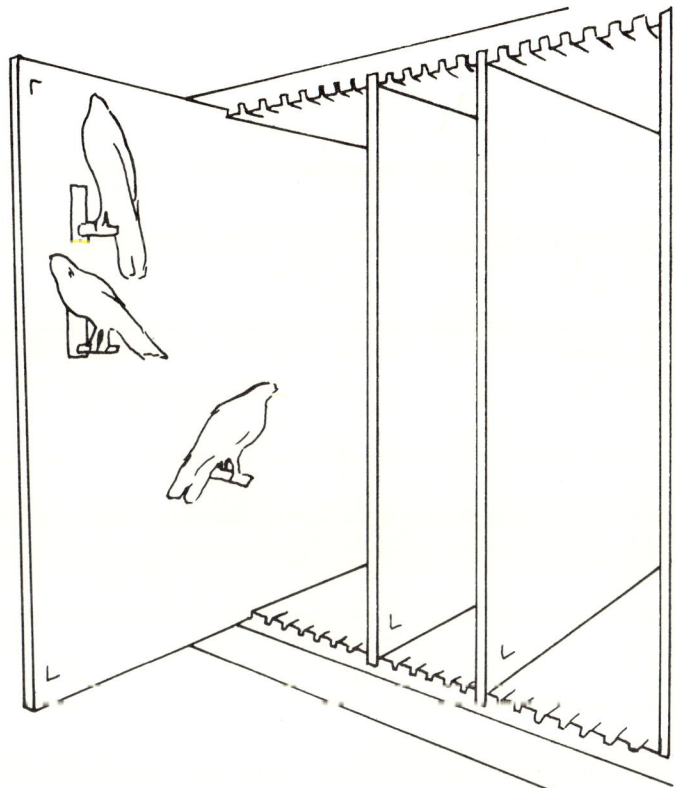

Fig. 21.—Panel system for storing mounted bird specimens (after Steel, 1970).

cause swelling and shrinkage. This stress may lead to deterioration and breakage of the tape (McWilliams, 1979).

Small collections which do not have adequate personnel or facilities for preserving and maintaining a sound recording collection may wish to consider sending the original tape, or a very high quality copy, to a larger, central collection. Not only does this permit proper preservation treatment, but it would make the data on the tape more accessible to more scientists (Gulledge, 1977, 1979; Hardy, 1984).

### Parasites and Stomach Contents

These preparations are generally stored in small glass vials or jars, with rubber stoppers or screw-top lids. Vials and jars may be stored in racks (Fig. 22) or in drawers of a metal cabinet (Fig. 23). The racks are inexpensive and easy to construct (McCafferty, 1974), but a larger collection may find the cabinet arrangement more efficient.

Stomach contents may be ordered in numerical sequence based on the catalog number of the bird specimen from which they are removed. Small

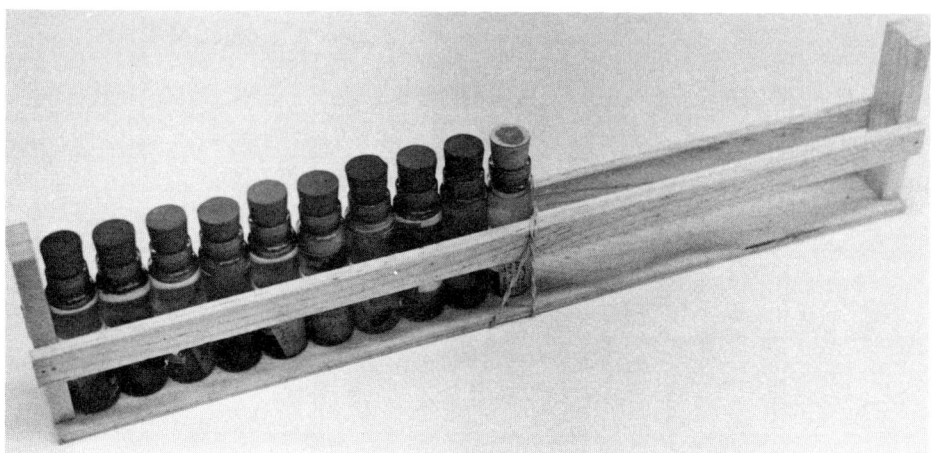

Fig. 22.—Racks for storing parasites or stomach contents.

parasite collections may ordered similarly, but it may be more efficient to organize larger collections phylogenetically.

### Slides and Photographs

As support material for ornithological studies or as a collection in itself, photographic materials pertaining to birds and their lives should receive the same care as that given to study skins.

Photographic materials are particularly vulnerable to deterioration. Haas (1983) summarized the basic conservation and storage needs for a collection of photographic material and provided a list of suppliers of conservation products. The six basic considerations detailed by Haas (1983) are: 1) control of atmospheric conditions, 2) control of dust and handling, 3) control of the chemistry of the storage and containment system, 4) "problem" materials (some require separate storage conditions), 5) cataloging, and 6) preparation of disaster and use plans. Collection managers should note that care should be taken even in choosing storage envelopes because some of the plastics that have been used commercially to make these envelopes speed up the deterioration process rather than retard it (Haas, 1983; Rempel, 1983).

As an alternative to financing and maintaining an extensive photographic collection, curators may wish to consider contributing their original material to VIREO. The Academy of Natural Sciences of Philadelphia "founded VIREO to be a centralized collection of bird photographs—to make them accessible for scientific and public use, to apply modern techniques in archival storage, and more, to develop new applications for photography within ornithology" (VIREO brochure). Copies of their acquisition policy and instructions for making contributions are available directly from VIREO (see Appendix II for the address).

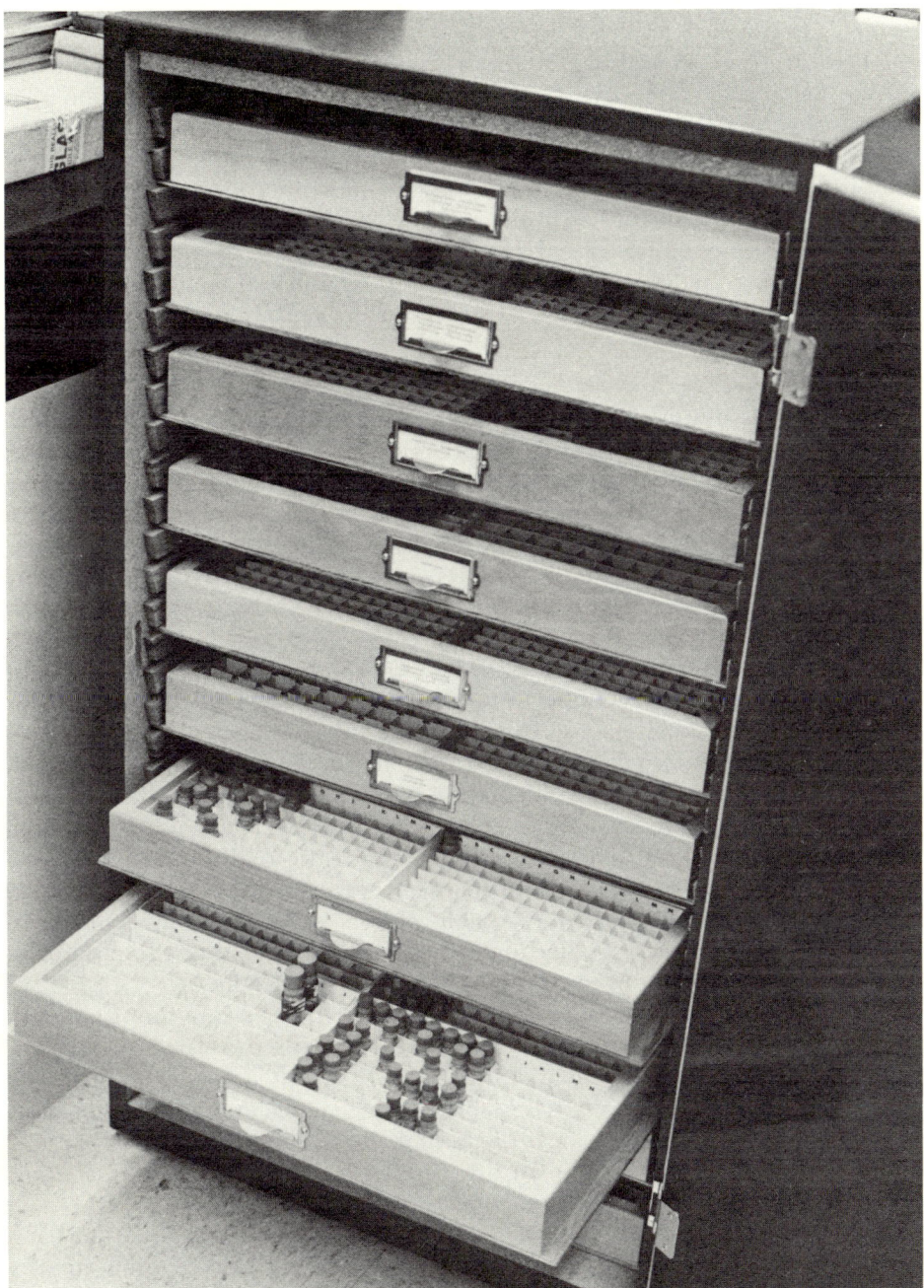

Fig. 23.—Storage of parasites in metal cabinet.

Field Notes

These original records of field work and collecting activities should be stored in a dust-free environment, protected from light and potential damage by insects, fire, and water. Access to and use of notes should be controlled in order to limit unnecessary wear. One method for limiting handling of the original notes is to make copies (microfilm, microfiche, or photocopies) available for use in place of the originals.

Field notes may be filed in archival folders and files, or more commonly, bound together in logical units; for example, notes of one collector over several years in one volume.

## Insurance

A collection should consider obtaining insurance against major losses and damages caused by such hazards as fire, floods, severe weather, vandalism, and theft. Policies should be considered for both the building housing the collection and the collection itself. "The insurance coverage on museum collections today is as varied as the collections themselves. These policies range from fine arts dealers' policies to manuscript policies of many descriptions, and all of them differ significantly" (Babcock and Haack, 1981).

Eighteen of the collections surveyed by the author reported insurance of their facilities, whereas only 13 reported insurance on the specimens themselves. Some institutions, though, have questioned the practicality of insuring certain parts of their collections, if not all. Because part of each specimen's inherent value is dependent on the time and place it was collected, a specimen is irreplaceable. Some species, such as extinct and endangered species, can never be collected again. In addition, the monetary value of zoological specimens is difficult to evaluate. Art museums also are faced with the problem of irreplaceable objects, and based on a survey of art museums' insurance, Pfeffer and Uhr (1974) learned that the premiums paid for insurance amount to 10 times the cost of insured losses.

When considering insurance policies, it is best to become familiar with the terminology used. Dudley et al. (1979) discussed various aspects of insuring collections and includes an excellent glossary of terms. The Association of Art Museums has published a detailed manual on fine arts insurance (Nauert and Black, 1979), and Babcock and Haack (1981, 1982) proposed a simplified, standardized policy for museum collections. Unfortunately, the latter is not yet widely used.

Although Lawton and Block (1966) focused primarily on fine arts collections, their summary of what to consider in a policy applies equally to bird collections. Of particular importance is the need to identify and insure both the "permanent collection" (those specimens which never leave the institution) and the "loan collection" (those specimens which are either loaned out or received for a temporary period). Both collections must be satisfactorily covered, but each generally requires a different coverage. It is generally best to insure only against major disasters, not minor problems.

## INFORMATION RETRIEVAL

One of the primary functions of collection management is to establish an efficient system to retrieve information and specimens from the collection. Assigning a catalog number to each specimen is a prerequisite for developing such a system. Specimens and their data may then be cross-referenced according to the projected needs of collection users. Generally, the most useful cross-reference files are based on taxonomic groups and geographic distinctions.

Cross-references may be created either manually or by an electronic data processing system. There are many factors to consider before a decision can be made between these two systems (Chenall, 1975; Humphrey and Clausen, 1977). For small, and many medium-sized, collections it may be more cost efficient in terms of staff time, and costs of equipment and supplies, to generate manual files rather than computerized files. The primary cost for manually generated files is the staff time required to maintain accurate files and to perform information searches from the completed files. For either a small collection or one that is growing very slowly, this cost will be minimal.

Computerized files are generally more expensive but frequently are more versatile than manual files. The institution must obtain use of the necessary computer hardware, and frequently hire a programmer to create the necessary software for the system. Staff time also is required to enter and edit data in the computer's master file. However, these costs need not be exorbitant (Sarasan, 1979); minicomputers, microcomputers, and prepackaged programs increase efficiency and decrease costs markedly (McAllister et al., 1978; Koeppl, 1981; Manning and Cunningham, 1982; McAllister, 1983; Cato and Folse, 1985).

A computerized catalog system can be used in ways a manual system cannot. If properly programmed, it can generate not only master catalogs based on numeric, taxonomic, and geographic units, but it can generate labels for skeleton boxes and vials as well. A system for maintaining loan and exchange records might also be included. Data corrections and taxonomic updating may easily be accomplished with editing programs. In addition, programs of data manipulation for specific research studies make many complicated comparisons possible that previously were too time-consuming to complete (James and Karr, 1973).

Of the collections surveyed by the author, 15 reported the use of a manual system only for retrieving data; three reported computer files only; eight reported using both; and two reported having no retrieval systems other than the numerical catalog. The collections also were questioned concerning 12 fields of information that might be used on which to base data requests: catalog number, accession number, A.O.U. number, order, family, genus, species, subspecies, country, state, county, and specific locality. Only one of the collections with manual files only could easily retrieve data for more than six of these fields, whereas the collections with computer files only had access to 8, 9, and 11 fields. The collections with both manual and computer files had access to 6, 7, 10, 11 and 12 fields.

The potential effectiveness and value of a computerized system should be determined for each collection based on its specific needs, resources, and priorities. General information concerning computerization of collections can be found in "Museums and Computers: Strange Bedfellows" (Museum News, 1973), the Association of Systematics Collections Newsletter (Association of Systematics Collections, 1975), Chenall (1975), and Sarasan and Neuner (1983). Chenall (1975) provided a detailed overview of the computerization process as well as descriptions of several systems developed by collections in American museums. Sarasan and Neuner (1983) presented a summary of the results of an Association of Systematics Collection survey concerning computer sytems in collections. This publication also discussed common problems encountered in computerization projects, and ways these could be avoided.

### Manual Files

The format for each reference file for specimen data should be standardized. The more uniform a file, the more accurate and useful it is likely to be. Many collections use index cards preprinted with the fields of information to be included (Figs. 24-25). To answer an information request, a photocopy of the card(s) is sent to the researcher.

In addition to the cross-reference system of specimen data, files should be maintained for accession forms, donation forms, loan invoices, exchange forms, and deaccession forms. A card file of donors and collectors listing addresses, telephone numbers, and type of donation may prove to be useful as well. Collecting permits also should be filed permanently in the collection area. Gallant (1981) provided practical advice on files management.

Additional documentary material, such as field notes, should be cross-referenced to provide better access. Ideally, general content is indexed as well as references to specific specimens. Archival practices provide procedures for organizing and indexing journals, manuscripts, photographs, and slides (Vanderbilt, 1966; Gilley, 1974; Bowditch, 1975; Cunha, 1975; Hommel, 1979; McKay, 1982). Original documentary material should not leave the collection area; researchers should be requested to use the material on-site.

### Computerized Files

If the computer is to be an effective retrieval tool, it is necessary to decide how the specimen data are to be used before any information is entered into the file. A request for data concerning county distributions can be answered only if data on counties are entered as a unique field of information. Therefore, it is necessary to determine during the planning phase which categories of information are to be included in the computer files.

| Reference Catalogue | | | | Texas A&M University | |
|---|---|---|---|---|---|
| | | *Fulmarus glacialis* | | Fulmar | |
| Dept. No. | Sex Age | Locality | | Date | Collector |
| 9825 | skeleton only M | HOLLAND:  Kijkduin, beach | | 27 Oct 1974 | E.Bouwman |
| 9826 | skeleton only F | "      :  's-Gravenzande, beach | | " | " |
| 9951 | F | CALIFORNIA:  Humboldt Co; Clam beach | | 17 Nov 1969 | D.L.Hook |
| 9986 | ? | CALIFORNIA:  Monterey Co;nr. Potrero Rd. Moss Landing | | 6 Mar 1976 | J.L.Cross |
| | | | | | |
| | | | | | |
| | | | | | |
| | | | | | |
| | | | | | |
| | | | | | |

Fig. 24.—Card from taxonomic cross-reference file, Texas Cooperative Wildlife Collection (original size: 4 by 6 inches).

| 2 | | | | | |
|---|---|---|---|---|---|
| **SPECIES** *Grallaria andicola* | | | | | |
| MUS. NO. | LOCALITY | DATE | COLLECTOR | SEX | NATURE |
| 80574 | Munte above (S) Yanac, ca. 13,000' Peru: Depto. Ancash; Bosque Quipis | 1 June 1975 | T. Parker | ♂ | skin |
| 80575 | tween Churubamba and Hda. Paty, above Acomayo ca. 11,000' " Depto. Huánuco; Unchog, pass be- | 12 June 1975 | Manuel Villar | ♂ | " |
| 81988 | ca. 10 km S Yanac, ca. 12,500' " Dpto. Ancash; Quebrado Tutapac, | 28 May 1976 | T. Parker | ♂ | " |
| 81989 | "        "        " | " | " | ♀ | " |
| 84957 | Central El. 11,800' " Dpto. Pasco; km 335 Carretera | 2 Aug. 1977 | G. R. Graves | ♀ | " |
| 91512 | Caldera, 7 km NE Tayabamba, 3550m " Dpto. La Libertad; Quebrada La | 10 Aug. 1979 | T.S.Schulenberg | | alcoholic |
| G *andicola* 92447 | pampa, 3350m " " 21.7 road km W Arica- | 28 July 1979 | J.William Eley | ♂ | skin |
| 92448 | 7 km NE Tayabamba, 3550m " " Quebrada La Caldera, | 8 Aug. 1979 | Mark B. Robbins | ♂ | " |
| 92449 | "        "        " | 9 Aug. 1979 | T.S.Schulenberg | ♂ | " |
| 92450 | "        "        " | 11 Aug. 1979 | J. William Eley | ♂ | " |
| 92451 | "        "        " | 15 Aug. 1979 | " | ♀ | " |
| **Museum of Zoology, Louisiana State University** | | | Bird Collection | | |

Fig. 25.—Card from taxonomic cross-reference file, Museum of Natural Science, Louisiana State University (original size: 5 by 8 inches).

CATEGORY:   FAMILY

DESCRIPTION:   This category applies to the most recent taxonomic designation of the family of the specimen.

FORMAT:   Enter the official spelling of the family as defined by the current rulings of the International Code of Zoological Nomenclature.

ACCEPTED VARIATIONS:   Depending on desired output and utilization, the families of mammals may be coded (see COMMENTS). If the family has not been determined, enter "UNKNOWN."

OMIT CONDITIONS:   If this category is adopted by the institution, it should not be omitted.

CONTINGENCY REQUIREMENTS:   None.

VALID EXAMPLES:   SOLENODONTIDAE
PHYLLOSTOMATIDAE
GEOMYIDAE
ANTILOCAPRIDAE
UNKNOWN

COMMENTS:   Utilization of this category is primarily for ease in retrieval and providing a general phylogenetic arrangement by grouping specimens at the familial level. The phylogenetic arrangement can be improved by special programming or by coding data. One disadvantage in coding data in this category is that input errors might be difficult to detect. If an institution desires to have coded information written out completely, such a function can be provided with special programming.

Fig. 26.—Example of data standardization for computer files (from Williams et al., 1979).

It also is necessary to standardize the data entered into the master file in order to generate accurate reference files. One main reason why computer cataloging projects do not meet expectations is that insufficient quality control is imposed on data as they enter (Sarasan, 1981). The American Society of Mammalogists' Committee for Information Retrieval has developed a guideline for data standards in Recent mammal collections (Williams et al., 1979). This serves as an excellent example of what is needed before data entry ever begins. Categories are listed and defined; the format is described and examples are given (Fig. 26). Other techniques that improve the ability to control data quality are 1) the use of an interactive computer system with a preformatted data screen that prompts personnel for each category of data that is to be entered (Fig. 27); and 2) programs for editing data in a systematic fashion.

Once data entry is complete, and with appropriate programming, it will be possible to generate printed cross-reference files as needed.

```
--------------------------------------------------------------------------------
=============================== Add Specimen ======================= 03-05-84 =
                                                                        ID: SN
                             Bird Division

    1.  Catalog #:..............(008267)
    2.  Order:......(   Podicipediformes)   11.  Sex:..............................(F)
    3.  Family:.......(   Podicipedidae)     12.  Gonad Condition:....................(5)
    4.  Genus:....(           Podiceps)      13.  Fat Condition:......................(4)
    5.  Species:..(         nigricollis)     14.  Skull Ossification:..............(  ?)
    6.  Subspec:..(         californicus)    15.  Special No.:................(         )
    7.  Date Coll:..........(11Nov1969)      16.  Specimen Condition:..............(sn)
    8.  Country:.......(           USA)      17.  Collector:......(          Coon,D.)
    9.  State:.........(         Texas)      18.  Preparer:.......(      Arnold,K.A.)
    10. County:.............( Chambers)      19.  Preparer's Number:...........(  3008)

    20. Specific Locality:......(                           6 mi S. Cove)
    21. Rem:(                     wt. 210.5g; ov. 11x5, l.o.<1; mallaphaga; shot)




    ===========================================■■■■■====--=================================
    <RET> accepts current entry, <new entry> replaces old.


    --------------------------------------------------------------------------------
```

Fig. 27.—Screen format for data entry; Texas Cooperative Wildlife Collection computerization project.

## COLLECTION USE

If collections are "to yield information, contribute to knowledge, or provide stimuli for aesthetic responses, they must be available for study and to view" (Force, 1975). Specimens in a collection may be used by members of both the scientific community and general public. The collections surveyed by the author reported use of their specimens by professional ornithologists (43.3%), students (23.3%), amateur ornithologists (11.8%), artists (10.4%), college teachers (3.2%), commercial users (1.9%), elementary and secondary school teachers (1.3%), and archaeologists and wildlife law enforcement officers (2.5%). These groups of individuals vary in their knowledge of the proper use of specimens, and it is the responsibility of the curator(s) and collection manager(s) to regulate the use of specimens in order to protect their integrity and prolong their life span.

Unlimited access is not necessarily desirable, and an attempt should be made to decide in advance who may use which specimens and under what conditions. A written access policy indicates that permission to use the collection is not given arbitrarily, but is based on guidelines designed to protect the specimens while maximizing their use for legitimate projects. It also establishes a standard policy in the event the curator and collection manager cannot be present to monitor collection use. Eleven of the collections surveyed by the author have written access policies, and eight reported having some form of user's guide.

Access policies should consider visits as well as loans to and from the collection, and in general, specimens from research collections should be available for use by responsible individuals associated with recognized educational or scientific organizations. Access to research specimens by ornithologists not associated with an organization should be left to the discretion of the curator and collection manager on a case-by-case basis. Specimens in teaching collections, plus less valuable mounted specimens, should be available for educational classes, artists, and commercial users. The host institution should retain the right, at the discretion of the curator and the director, to refuse admittance to anyone not conducting himself in a professional manner. The safety and integrity of the specimens always must be considered first.

*Visits*

Arrangements to use a collection generally should be made in advance by telephone or letter so that a visit may be as convenient as possible for both the visitor and the host. The host institution should provide staff assistance and work space for the visitor whenever possible. Special arrangements should be approved by the curator.

A staff member should acquaint a visitor with the organization of the collection and procedures for using specimens. If the visitor is unfamiliar with collection procedures, it is the responsibility of the staff member to explain how specimens should be handled. Brightly colored "specimen removal tags" (Fig. 28) should be easily available to mark the location of specimens removed for study.

Some collections provide users with a set of written guidelines including such information as a list of personnel and telephone numbers, a floor plan of the collection area, work space, offices, a list of available data sources other than the specimens, and procedures for specimen use and loans (Fisher, 1978). Such guides benefit not only the visitor, but the collection as well. Specific collection management procedures vary from one institution to another, but written guidelines omit the need for guessing on the part of the visitor.

Special attention should be given to requests to use type specimens and specimens of extinct and endangered species. Their unique status should limit their use, and it should be necessary to obtain permission directly

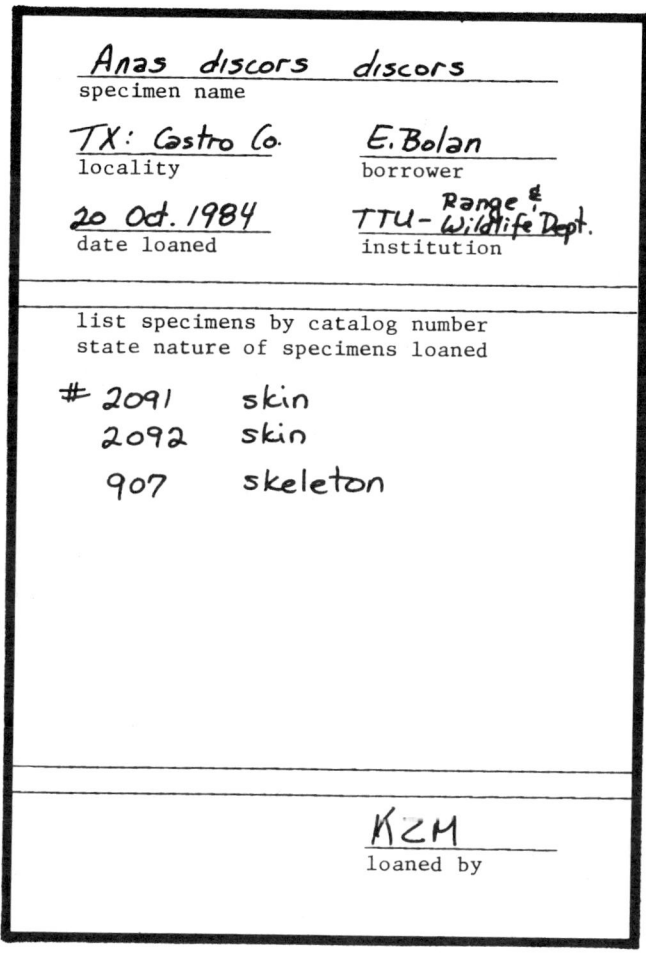

Fig. 28.—Specimen removal tag used at The Museum, Texas Tech University (color: blue; size: 4 by 6 inches).

from the curator for the use of those specimens. Particular care in handling must be taken with those specimens.

### Loans

Procedures for loans also should be established to protect the specimens. Written guidelines permit standardization of procedures and serve as a reference for part-time and temporary personnel. Thirteen of the collections surveyed by the author have a written loan policy.

### Outgoing Loans

All loan requests should be written and directed to the curator, stating the material requested and its intended use. Students should have loan requests cosigned either by their major professor or the curator of the collection to which the loan is to be sent. Each request should be reviewed individually and approved at the discretion of the curator.

Twelve of the collections surveyed by the author prefer to make loans only to an institution; four, to individuals at institutions; four, to individuals; five, to institutions or individuals. A loan made to an institution rather than an individual makes the institution responsible for the safety of the specimens. This is particularly important should the specimens be damaged while on loan or should the individual be delinquent in returning a loan and change employment without returning borrowed specimens.

If it is practical for the borrower, the curator should request that the specimens be used in the collection to decrease the possibility of specimen damage due to transportation. In addition, certain specimens should never leave the collection area, including specimens of extinct species and type specimens. Twenty of the collections surveyed reported that types were never loaned; five reported loaning them on rare occasions; and one responded that types were loaned on request. As for state records, 19 of the collections reported that they are loaned; one, rarely; seven, not at all.

Federal permits are required for interstate shipping of endangered species that were not collected and housed in a museum before 28 December 1973 (Genoways and Choate, 1976). The same is true for Bald Eagles acquired after 8 June 1940 and Golden Eagles acquired after 24 October 1962. Shipping containers must be labeled on the outside with the name and address of both the shipper and the consignee, as well as with an accurate statement of the contents. The latter requirement may be met by attaching a copy of the loan invoice in an envelope on the outside of the package (Genoways and Choate, 1976).

Loan material should be stored in dust-proof, fumigated cabinets, and it should not be altered in any way by the borrower without specific approval by the curator. Loans should not be made to individuals who either cannot or will not meet those conditions.

Most institutions do not charge researchers for use of collection specimens. However, seven of the collections surveyed by the author reported assessing fees for specimen use; five of the seven stated that the fees were primarily for artists and commercial users. The lending institution generally pays for insurance and shipping charges on outgoing material, expecting the borrower to pay the return costs. Only one institution of those surveyed specifically requests the borrower to insure the specimens while they are on loan; three of those which said they did not actually request insurance reported that they assumed the specimens would be safe.

In most cases, it is best to limit the size of a loan to any one borrower. If necessary, the specimens may be sent in two or three shipments, each contingent on the return of the first. When possible, at least one representative of a species should be kept in the collection at all times.

Loans should be prepared and packaged under the direction of a staff member. When a specimen is removed from its location, it should be replaced by a "specimen removal tag" (Fig. 28). These tags simplify the replacement of specimens when they return to the collection; the tags also immediately identify the location of the specimen to other potential users.

Loans should be documented fully (Fig. 29). A loan form should include the name and address of the borrower and the institution responsible for the safety of the specimens, the date of the transaction, the name of the individuals who approved and packaged the loan, the duration of the loan, and a detailed listing of the loan specimens. Most loans are made for six months, with requests for extensions handled by the curator. Half of the collections surveyed reported an average length of six months; the others responded with times of one month through one year.

The detailed listing of loan specimens should include for each, the catalog number, the type of preparation, and a description of its physical condition. The latter is particularly important; if specimens are returned with broken feet and missing feathers, it is impossible to know where the damage occurred if the physical conditions had not been noted when the specimens left the host institution. They may have been in poor condition before they were loaned, or damage may have resulted from transportation. However, damage may also have been caused by careless handling on the borrower's part. If the specimens have not been professionally cared for, loans to that particular borrower should not be repeated.

## Packing Loans

Loans should be packed in sturdy wooden boxes, with sufficient packing material to prevent movement of the specimens. Box tops fastened with screws instead of nails make repeated use of each box more practical.

Each specimen should be wrapped with tissue paper or cotton, or placed in cylinders made from light-weight cardboard (Wythe, 1938). Extra care should be taken to be certain that bills, feet and feather tips are protected fully. Skeletal preparations should be padded inside their boxes or vials in order to prevent fragmentation of the pieces. Alcoholic specimens should be wrapped in cheesecloth that has been dampened with alcohol, then enclosed with a small amount of alcohol inside a plastic bag. This prevents dehydration of the specimens during shipping. It is advisable to seal the first bag inside a second one in order to prevent leakage.

An address label and loan invoice should be packaged with the specimens inside the box. If the outside labels are lost, the material can still reach its destination. The address label on the outside of the package should include the full address of the lender, the borrower, and the value of the specimens for insurance purposes. Another invoice inside an envelope taped to the outside of the package fulfills legal requirements for interstate transportation as regulated by the Lacey Act of 1903. The following information should be included on the address label: BIOLOGICAL SPECIMENS, NO COMMERCIAL VALUE, and when applicable, NO ENDANGERED SPECIES.

INVOICE OF SPECIMENS

LOUISIANA STATE UNIVERSITY MUSEUM OF ZOOLOGY
Baton Rouge, Louisiana, U. S. A. 70893

Division of _____        Invoice Number _____
To _____                 Packed By _____
   _____                  Approved _____
   _____                  Date Shipped _____
   _____                  Shipped By _____
   _____                  Insured For _____

         The material below contained in _____ is transmitted as:

( ) a gift                    ( ) in exchange                ( ) return of borrowed material

( ) a loan at your request    ( ) for examination at our request

Please sign and return the blue copy to:    Received in good condition _____
     Museum of Zoology                                                      (date)
     Drawer MU, LSU
     Baton Rouge LA  70893                   Signature of recipient _____

         Loans are ordinarily made for a period of six months or less.
    A request for renewal should be made if specimens are needed for a longer period.

Fig. 29.—Invoice of specimens used at the Museum of Natural Science, Louisiana State University (4-part NCR paper; original size: 8½ by 11 inches).

## Incoming Loans

Procedures for incoming loans should be similar to those for outgoing loans. An institution should treat borrowed specimens the same way it would like its own specimens to be treated. In order to ensure the proper handling of borrowed materials, it might be advantageous to have a short set of written guidelines available for researchers who wish to house loans in the collection.

Specimens are requested for short-term loans (up to six months) only. Responsible individuals and researchers affiliated with other institutions should receive permission from the curator to receive and store loan specimens in the collection. All materials borrowed should be stored in fumigated cabinets separately from the research and teaching collections. If requested by the lending institution, the borrower will pay for any insurance or shipping costs. Written agreement of financial responsibilities should be completed before the loan is processed.

The specimens should be inventoried on arrival, and the physical condition noted on the invoice sent by the lending institution. If such an invoice is not included, the receiving institution should use its own form. The lending institution should be notified of the receipt and physical condition of the material.

Specimens should be returned in the same manner as they arrived. They are again inventoried and their physical condition checked before being packaged for return. When possible, the specimens should be returned in the same box. The lending institution should be notified by letter that the loan is being returned. A copy of the loan invoice should be enclosed with the loan. One method for verifying the safe return of the borrowed material is to enclose a self-addressed and stamped postcard with a request that a representative of the lending institution sign and return the card upon the satisfactory receipt of the specimens.

## Exhibit Specimens

Specimens loaned for exhibit purposes should be documented in the same manner as study skins. However, because they are more likely to be kept in conditions that are less than ideal by conservation standards, they should be accompanied by guidelines for use. Particularly important are standards for handling, lighting, insect protection, and security. Mounted specimens have a more obvious appeal to the general public and, therefore, are more susceptible to theft and vandalism.

## CONCLUSION

Efficient collection management has become an essential component in the operation of ornithology collections. Although numerous references deal with various aspects of collection management, very few tie these phases together. This publication is an attempt to provide some basic considerations for collection management as based on information in the literature or practices used in North American collections. Through the use of professional management practices, the quality and value of ornithology collections may be increased.

## Acknowledgments

I would like to extend my sincere thanks to the curators and collection managers who responded to the questionnaire: J. P. Angle, J. C. Barlow, G. F. Barrowclough, R. G. Bauer, P. Brodkorb, D. M. Devaney, J. W. Fitzpatrick, F. B. Gill, J. Hafner, J. W. Hardy, T. R. Huals, N. K. Johnson, R. E. Johnson, L. F. Kiff, M. J. Mengel, D. M. Niles, J. R. Northern, H. Ouellet, O. T. Owre, K. C. Parkes, R. A. Paynter, Jr., E. E. Pillaert, A. Rea, J. V. Remsen, Jr., S. A. Rohwer, S. M. Russell, R. W. Schreiber, C. G. Sibley, R. W. Storer, G. Wine, and R. L. Zusi. Thanks also must be extended to K. Z. Marshall for work contributed to an earlier version of this manuscript; to B. J. Fisher and N. L. Olson for photographic assistance; and to D. James, M. Le Croy, D. Maurer, M. B. Robbins, K. Russell, and T. Webber for information on specific topics. M. J. Mengel and D. J. Schmidly provided encouragement and numerous suggestions, and I am particularly grateful to K. A. Arnold, C. Jones, L. F. Kiff, and J. V. Remsen, Jr., for reviewing the manuscript and providing valuable suggestions.

## Literature Cited

Allen, J. F. 1974. Systematics collections—an essential resource in environmental assessment. Assoc. Syst. Coll. Newsletter, 2(3):4.

American Association of Museums. 1978. Museum ethics. Amer. Assoc. Mus., Washington, D.C., 31 pp.

American Association of Museums Energy Workshop Planning Committee. 1977. Protection of collections during energy emergencies. Amer. Assoc. Mus., Washington, D.C., 19 pp.

American Museum of Natural History. 1974. A statement of policy and procedures regulating the acquisition and disposition of natural history specimens. Curator, 17:83-90.

American Ornithologists' Union. 1957. Check-list of North American birds. 5th ed. Amer. Ornith. Union, Lord Baltimore Press, Baltimore, 691 pp.

———. 1975. Report of the ad hoc committee on scientific and educational use of wild birds. Auk, 92(3):1A-27A.

———. 1983. Check-list of North American birds. 6th ed. Amer. Ornith. Union, Washington, D.C., 877 pp.

Anderson, R. M. 1965. Methods of collecting and preserving vertebrate animals. Bull. Nat. Mus. Canada, 69:vii+1-199.

Anderson, S. 1965. Sources of error in locality data. Syst. Zool., 14:344-346.

Association of Systematics Collections. 1975. Computerized cataloging; basic considerations. Assoc. Syst. Coll. Newsletter, 3(5):9-11.

———. 1976. Compactors: one solution to the problem of collection growth. Assoc. Syst. Coll. Newsletter, 4:20-22.

———. 1979. User's guide to the Center for Biosystematics Resources. Assoc. Syst. Coll., Lawrence, Kansas.

———. 1981. The role of the ASC in strengthening university systematics collections. Assoc. Syst. Coll. Newsletter, 9(1):1-4.

Avise, J. C., and R. L. Crawford. 1981. A matter of lights and death. Nat. Hist., 90(9):6-14.

Axtell, R. W. 1965. More on locality data and its presentation. Syst. Zool., 14:64-68.

Babcock, P. H., and M. T. Haack. 1981. Plain-English collections insurance. Mus. News, 59(7):22-25.

————. 1982. A new easy to understand insurance policy for museum collections. Smithsonian Institution Press, Washington, D.C.

BANKS, R. C. (ED.). 1979. Museum studies and wildlife management: selected papers. Smithsonian Institution Press, Washington, D.C., 297 pp.

BANKS, R. C., M. H. CLENCH, AND J. C. BARLOW. 1973. Bird collections in the United States and Canada. Auk, 90:136-170.

BARLOW, J. C., M. H. CLENCH, J. L. GULLEGE, J. R. JEHL, N. K. JOHNSON, AND R. L. ZUSI. 1977. AOU/NSF Workshop on a National Plan for Ornithology. Draft report. Amer. Ornith. Union.

BARLOW, J. C., AND N. J. FLOOD. 1983. Research collections in ornithology—a reaffirmation. Pp. 37-54, *in* Perspectives in ornithology (A. H. Brush and G. A. Clark, Jr., eds.). Cambridge Univ. Press, New York.

BEAN, M. J. 1983. The evolution of national wildlife law. Rev. ed. Praeger, New York, 448 pp.

BERGER, A. J. 1955. Suggestions regarding alcoholic specimens and skeletons of birds. Auk, 72:300-303.

————. 1956. Further notes on alcoholic specimens. Auk, 73:452.

BERGER, T. J., AND J. D. PHILLIPS. 1977. Index to U.S. Federal Wildlife Regulations (with quarterly updates). Assoc. Syst. Coll., Lawrence, Kansas.

————. 1981. Directory of state protected species: a reference to species controlled by nongame regulations. Assoc. Syst. Coll., Lawrence, Kansas.

BERGER, T. J., A. M. NEUNER, AND S. R. EDWARDS. 1979. Directory of federally controlled species. Assoc. Syst. Coll., Lawrence Kansas.

BORROR, D. J., D. M. DELONG, AND C. A. TRIPLEHORN. 1981. An introduction to the study of insects. 5th ed. Saunders College Publishing Co., Philadelphia, 827 pp.

BOWDITCH, G. 1975. Cataloging photographs: a procedure for small museums. Tech. Leaf. Amer. Assoc. State Local Hist., 57:1-4.

BUCK, R. D. 1964. A specification for museum air-conditioning. Mus. News, 43(4):53-57.

CAMERON, D. 1968. Environmental control: a theoretical solution. Mus. News, 46(9):17-21.

CATO, P. S., AND L. J. FOLSE. 1985. A microcomputer/mainframe hybrid system for computerizing specimen data. Curator, 28:105-116.

CHENHALL, R. G. 1975. Museum cataloging in the computer age. Amer. Assoc. State Local Hist., Nashville, 261 pp.

CLENCH, M. H., R. C. BANKS, AND J. C. BARLOW. 1976. Bird collections in the United States and Canada: addenda and corrigenda. Auk, 93:126-129.

CRAWFORD, R. L. 1983. Grid systems for recording specimen collection localities in North America. Syst. Zool., 32(4):389-402.

CUNHA, G. M. 1975. Conserving local archival materials on a limited budget. Tech. Leaf. Amer. Assoc. State Local Hist., 86:1-4.

DENVER MUSEUM OF NATURAL HISTORY. 1978. Collection policies and procedures. Denver Mus. Nat. Hist., Denver, 31 pp.

DESSAUER, H. C., AND M. S. HAFNER (EDS.). 1984. Collections of frozen tissues. Assoc. Syst. Coll., Lawrence, Kansas, 74 pp.

DOWLER, R. C., AND H. H. GENOWAYS. 1976. Supplies and suppliers for vertebrate collections. Museology, 4:1-83.

DUDLEY, D. H., S. B. WILKINSON, AND OTHERS. 1979. Museum registration methods. 3rd ed., revised. Amer. Assoc. Mus., Washington, D.C., 437 pp.

EDWARDS, E. P. 1974. A coded list of birds of the world (edition a). Sweet Briar, Va.

EDWARDS, S. R., B. M. BELL, AND M. E. KING. 1981. Pest control in museums: a status report (1980). Assoc. Syst. Coll., Lawrence, Kansas.

EDWARDS, S. R., AND L. D. GROTTA (EDS.). 1976. Systematics collections and the law. Assoc. Syst. Coll., Lawrence, Kansas, 27 pp.

FELLER, R. L. 1964. The deteriorating effect of light on museum objects. Tech. Suppl. 3, Mus. News, Amer. Assoc. Mus., viii pp.

————. 1968. Control of deteriorating effects of light on museum objects. Mus. News, 46(9):39-47.

FIELD MUSEUM OF NATURAL HISTORY. 1976. Policy statement on accessions and deaccessions. Field Museum of Natural History, Chicago, 16 pp.

FINK, W. L., K. E. HARTEL, W. G. SAUL, E. M. KOON, AND E. O. WILEY. 1978. A report on current supplies and practices used in curation of ichthyological collections. Amer. Soc. Ichthyologists and Herpetologists, 63 pp.

FISHER, R. D. 1978. User's guide to the mammal collections. National Museum of Natural History, Smithsonian Institution, Washington, D.C., 19 pp.

FORCE, R. W. 1975. Museum collections—access, use, and control. Curator, 18(4):249-255.

FRITTS, T. H. 1976. Criteria for accession—one solution to the problem of collection growth. Assoc. Syst. Coll. Newsletter, 4(4):54-55.

GALLANT, W. 1981. Files inventory and reorganization. Infor. Rec. Manag., April:20-22.

GENOWAYS, H. H., AND J. R. CHOATE. 1976. Federal regulations pertaining to collection, import, export, and transport of scientific specimens of mammals. J. Mamm., 57(2, suppl.):1-9.

GILLEY, B. L. 1974. Declassifying the slide secrets. Mus. News, 52(6):45-48.

GULLEDGE, J. L. 1977. Recording bird sounds. Living Bird, 15:183-203.

————. 1979. The Library of Natural Sounds at the Laboratory of Ornithology, Cornell University. Recorded Sound, 74-75:38-41.

GUTHE, C. E. 1970. Documenting collections: museum registration and records. Tech. Leaf. Amer. Assoc. State Local Hist., 11:1-8.

HAAS, P. 1983. The conservation of photographic collections. Curator, 26(2):89-106.

HALL, E. R. 1962. Collecting and preparing study specimens of vertebrates. Misc. Pub. 30:1-46. Mus. Nat. Hist., Univ. Kansas.

HALL, E. R., AND W. C. RUSSELL. 1933. Dermestid beetles as an aid in cleaning bones. J. Mamm., 14:372-374.

HARDY, J. W. 1978. Suggestions for preparation of master tape recordings for production of phonodiscs for publication. Amer. Birds, 32(5):965-967.

————. 1984. Depositing sound specimens. Auk, 101(3):623-624.

HARRISON, C. J. O., AND G. S. COWLES. 1970. Birds. Instructions for collectors, No. 2A. British Mus. (Nat. Hist.), London, 47 pp.

HART, C. W., JR. 1978. The burden of regulation. Mus. News, 56(3).

HILDEBRAND, M. 1968. Anatomical preparations. Univ. California Press, Berkeley.

HOMMEL, C. 1979. A model museum archives. Mus. News, 58(2):62-69.

HOWER, R. O. 1970. Advances in freeze-dry preservation of biological specimens. Curator, 13(2):135-152.

————. 1979. Freeze-drying of biological specimens: a laboratory manual. Smithsonian Institution Press, Washington, D.C., 196 pp.

HUBNER, M. (ED.). 1981. A lesson on lighting. History News, 36(2):45-47.

HUMPHREY, P. S. 1972. A new organization: The Association of Systematics Collections. Curator, 15:32.

HUMPHREY, P. S., AND A. C. CLAUSEN. 1977. Automated cataloging for museum collections: a model for decision and a guide for implementation. Assoc. Syst. Coll., Lawrence, Kansas, 79 pp.

HUNTER, J. E. 1974. Preservation of objects in museum exhibits. Reprint from a session at Mountain-Plains Mus. Conf. annual meeting, Abilene, Kansas, 47 pp.

————. 1979. Emergency preparedness for museums, historic sites, and archives: an annotated bibliography. Tech. Leaf. Amer. Assoc. State Local Hist., 114:1-4.

————. 1980. Preparing a museum disaster plan, photocopy. U.S. National Park Service, Washington, D.C., 8 pp.

HUTCHINSON, U. H. 1964. Distance and direction in locality data. Syst. Zool., 13:158-159.

INTERNATIONAL COUNCIL OF MUSEUMS. 1970a. Ethics of acquisition. ICOM News, 23:49-58.

JAMES, F. C., AND J. R. KARR. 1973. Are computers going to the birds? Mus. News, 51(8):13-15.

JENKINSON, M. A., AND D. S. WOOD. 1985. Avian anatomical specimens: a geographic analysis of needs. Auk, 102(3):587-599.

JOHN, D. K. 1979. Information on the design and furnishing of the osteological preparation laboratory of the Smithsonian Institution. Museum of Natural History, Smithsonian Institution, Washington, D.C.

JOHNSON, N. K., R. M. ZINK, G. F. BARROWCLOUGH, AND J. A. MARTIN. 1984. Suggested techniques for modern avian systematics. Wilson Bull., 96:543-560.

KANNEMEYER, S. 1973. The storage of a wet collection. South African Mus. Assoc. Bull. 10(7):274-277.

KEAST, A. 1973. The role of the museum in ornithology. Emu, 73:242-247.

KENNEDY, R. A. 1960. Conservation in the humid tropical zone. Mus. News, 38(7):16-20.

KIFF, L. F. 1978. Instructions for the preparation and shipment of eggshell specimens. Western Foundation of Vertebrate Zoology.

———. 1979. Bird egg collections of North America. Auk, 96:746-755.

KING, J. R., AND W. J. BOCK. 1978. Workshop on a National Plan for Ornithology: final report, submitted to the National Science Foundation and the Council of the American Ornithologists' Union, 300 pp.

KNUDSEN, J. W. 1972. Collecting and preserving plants and animals. Harper and Row, New York, 320 pp.

KOEPPL, J. W. 1981. DMS: a simple, efficient computerized data management system for museums. Assoc. Syst. Coll. Newsletter, 9(2):19-21.

KOHL, M. F. 1982. Archival supplies for the small museum. History News, 37(1):34-36.

LAWTON, J. B., AND H. T. BLOCK. 1966. Museum insurance. Curator, 9:289-297.

LEE, W. L., B. M. BELL, AND J. F. SUTTON. 1982. Guidelines for acquisition and management of biological specimens. Assoc. Syst. Coll., Lawrence, Kansas, 42 pp.

MALARO, M. C. 1979. Collections management policies. Mus. News, 58(2):57-61.

MANNING, A., AND R. CUNNINGHAM. 1982. A parsimonious computer application for museum accession and loan management. Assoc. Syst. Coll. Newsletter, 10(1):5-7.

MAYR, E., AND R. GOODWIN. 1956. Biological materials. Part I. Preserved materials and museum collections. Nat. Acad. Sci. and Nat. Res. Counc., Washington, D.C., 399:1-19.

MCALLISTER, D. E. 1983. An introduction to minicomputers in museums. Pp. 105-136, in Proceedings of 1981 Workshop on Care and Maintenance of Natural History Collections (D. J. Faber, ed.). Nat. Mus. Canada, Syllogeus, 44:1-196.

MCALLISTER, D. E., R. MURPHY, AND J. MORRISON. 1978. The complete minicomputer cataloging and research system for a museum. Cuator, 21:63-91.

MCCABE, T. T. 1943. An aspect of collectors' technique. Auk, 60:550-558.

MCCAFFERTY, W. P. 1974. An economical and efficient vial rack for the storage of small fluid-preserved specimens. Ann. Ento. Soc. Amer., 67:996-997.

MCGAUGH, M. H., AND H. H. GENOWAYS. 1976. State laws as they pertain to scientific collecting permits. Museology, 2:1-81.

MCKAY, E. 1982. Guidelines for organizing a manuscript collection. History News, 37(4):44+.

MCWILLIAMS, J. 1979. The preservation and restoration of sound recordings. Amer. Assoc. State Local Hist., Nashville, 138 pp.

MERYMAN, H. T. 1960. The preparation of biological museum specimens by freeze-drying. Curator, 3:5-19.

———. 1961. The preparation of biological museum specimens by freeze-drying: II. Instrumentation. Curator, 4:153-174.

MORONY, J. J., W. J. BOCK, AND J. FARRAND, JR. 1975. Reference list of the birds of the world. Spec. Publ. Amer. Mus. Nat. Hist., New York.

MUSEUM NEWS. 1973. Museums and computers: strange bedfellows. Amer. Assoc. Mus., 51(8):1-48.

NATIONAL MUSEUMS OF CANADA. 1983. Collections policy and procedures. Codex musealis, 2:1-95.

NAUERT, P., AND C. M. BLACK. 1979. Fine arts insurance: a handbook for art museums. Assoc. Art Mus., Washington, D.C.

NICHOLSON, T. D. 1974. NYSAM policy on the acquisition and dispositition of collection materials. Curator, 17:5-9.

———. 1975. The Australian Museum and the Field Museum adopt policy statements regarding collections. Curator, 18:296-314.

NORRIS, R. A. 1961. A new method of preserving bird specimens. Auk, 78:436-440.

PALMER, W. M. 1974. Inexpensive jars for museum specimens. Curator, 17:321-324.

PARKES, K. C. 1963. The contribution of museum collections to knowledge of the living bird. Living Bird, 2:121-130.

PETERS, J. L., AND OTHERS. 1934-1979. Check-list of birds of the world. 15 vols. Museum of Comparative Zoology, Cambridge.

PFEFFER, I., AND G. B. UHR. 1974. The truth about art museum insurance. Mus. News, 52:23-31.

PHELAN, M. 1982. Museums and the law. Amer. Assoc. State Local Hist., Nashville, 287 pp.

PLENDERLEITH, H. J., AND P. PHILIPPOT. 1960. Climatology and conservation in museums. Museum, 13:242-289.

PRITCHARD, M. H., AND G. O. W. KRUSE. 1982. The collection and preservation of animal parasites. Univ. Nebraska Press, Lincoln, 141 pp.

QUAY, W. B. 1974. Bird and mammal specimens in fluid-objectives and methods. Curator, 17:91-104.

RATH, F. L., JR., AND M. R. O'CONNELL (eds.). 1977. Care and conservation of collections. Amer. Assoc. State Local Hist., Nashville, 107 pp.

REMPEL, S. 1983. Enclosures for housing photographic negatives. Conservation Notes, Materials Conserv. Lab., Texas Memorial Mus., Univ. Texas (Austin), 3:1-4.

REMSEN, J. V. 1984. Procedure for filling out LSUMZ bird labels, photocopied. Louisiana State University.

RICKLEFS, R. E. 1980. Old specimens and new directions: the museum tradition in contemporary ornithology. Auk, 97:206-207.

RIEMER, W. J. 1954. Formulation of locality data. Syst. Zool., 3:138-140.

RUSSELL, W. C. 1947. Biology of the dermestid beetle with reference to skull cleaning. J. Mamm., 28:284-287.

SARASAN, L. 1979. An economical approach to computerization. Mus. News, 57(4):61-64.

———. 1981. Why museum computer projects fail. Mus. News, 59(4):40-49.

SARASAN, L., AND A. M. NEUNER. 1983. Museum collections and computers. Assoc. Syst. Coll., Lawrence, Kansas, 292 pp.

SCHMIDT, R. H. 1972. How to mount birds. Kansas State Teachers College, Emporia, Kansas, 80 pp.

SIMMS, E. 1979. Wildlife sounds and their recording. P. Elek, London, 144 pp.

SMITHSONIAN INSTITUTION. 1957. Suggestions for collecting and preserving specimens of birds and their eggs. SIL, 123:1-11.

SOMMER, H. G., AND S. ANDERSON. 1974. Cleaning skeletons with dermestid beetles—two refinements in the method. Curator, 17:290-298.

STEEL, C. A. B. 1970. A system for the storage of mounted birds. Museums J., 70:10-12.

STOLOW, N. 1966a. The action of environment on museum objects. Part I: humidity, temperature, atmospheric pollution. Curator, 9:175-185.

———. 1966b. The action of environment on museum objects. Part II: light. Curator, 9:298-306.

———. 1977. Notes on the measurement of relative humidity and temperature for museums. AAM Energy Workshop Planning Committee, Amer. Assoc. Mus., Washington, D.C.

STORER, R. W. 1971. Classification of birds. Pp. 1-18 in Avian biology 1 (D. S. Farner, J. R. King, and K. C. Parkes, eds.). Academic Press, London, 586 pp.

TIEMEIR, O. W. 1940. The dermestid method of cleaning skeletons. Kansas Univ. Sci. Bull. 26:377-383.

TRUEB, L., AND S. R. EDWARDS. 1978. The Association of Systematics Collections. Mus. News, 56(3):12-15.

ULLBERG, A. D., AND P. ULLBERG. 1974. A proposed curatorial code of ethics. Mus. News, 52(8):18-22.

VALCARCEL, A., AND D. L. JOHNSON. 1981. A new dermestid repository for skeletal preparation. Curator, 24:261-264.

VANDERBILT, P. 1966. Filing your photographs: some basic procedures. Tech. Leaf. Amer. Assoc. State Local Hist., 36:1-8.

VAN GELDER. R. G. 1965. "Another man's poison." Curator, 8:55-71.

VAN TYNE, J. 1952. Principles and practices in collecting and taxonomic work. Auk, 69:27-33.

WAKE, D. B. 1975. Report of the Committee on Resources in Herpetology. Copeia, 1975:391-404.

WATSON, G. E., AND A. B. AMERSON, JR. 1967. Instructions for collecting bird parasites. Information Leaflet 477. Smithsonian Institution, Washington, D.C.

WETMORE, A. 1960. A classification for the birds of the world. Misc. Coll., Smithsonian Institution, Washington, D.C., 139(11).

WILLIAMS, S. L., R. LAUBACH, AND H. H. GENOWAYS. 1977. A guide to the management of Recent mammal collections. Spec. Pub., Carnegie Mus. Nat. Hist., 4:1-105.

WILLIAMS, S. L., M. J. SMOLEN, AND A. A. BRIGIDA. 1979. Documentation standards for automatic data processing in mammalogy. The Museum, Texas Tech University.

WOOD, D. S., R. L. ZUSI, AND M. A. JENKINSON. 1982a. World inventory of avian skeletal specimens. Amer. Ornith. Union and Oklahoma Biol. Survey, Norman, 224 pp.

———. 1982b. World inventory of avian spirit specimens. Amer. Ornith. Union and Oklahoma Biol. Survey, Norman, 181 pp.

WYTHE, M. W. 1938. Safe packing of dry study-skins of birds for shipment. Condor, 40:42-43.

ZUSI, R. L. 1969. The role of museum collections in ornithological research. Proc. Biol. Soc. Washington, 82:651-661.

ZUSI, R. L., D. S. WOOD, AND M. A. JENKINSON. 1982. Remarks on a world-wide inventory of avian anatomical specimens. Auk, 99:740-757.

ZWEIFEL, R. G. 1966. Guidelines for the care of a herpetological collection. Curator, 9:24-35.

APPENDIX I.—Bird collections and numbers of specimens reported in 1983 survey.

| Collection | Study Skins | Skeletons | Fluid | Clutches of Eggs | Nests | Mounted | Other |
|---|---|---|---|---|---|---|---|
| American Museum of Natural History New York, NY | 1,000,000 | 9,000 | 9,000 | 18,000 | 5,000 | thousands | |
| National Museum of Natural History Smithsonian Institution Washington, D.C. | 450,000 | 35,731 | 21,778 | 46,000 | 10,000 | 3,000 | C |
| Museum of Comparative Zoology Harvard University Cambridge, MA | 330,000 | 7,000 | ? | many | many | 5,000 | |
| Field Museum of Natural History Chicago, IL | 320,000 | 7,000 | 5,000 | 17,000 | 600 | 1,500 | |
| Museum of Vertebrate Zoology University of California Berkeley, CA | 185,000 | 20,000 | 4,700 | 9,000 | 4,000 | | FT |
| Academy of Natural Sciences Philadelphia, PA | 170,000 | | | | | 2,500 | |
| Carnegie Museum of Natural History Pittsburgh, PA | 160,000 | 7,500 | 3,500 | 9,700 | several hundred | 1,000 | SW |
| University of Michigan Museum of Zoology Ann Arbor, MI | 150,000 | 18,000 | several hundred | several hundred | some | | SR |
| Royal Ontario Museum Toronto, Ontario | 120,000 | 28,000 | 4,400 | 11,500 | 3,000 | 400 | SR |
| Museum of Vertebrate Zoology Louisiana State University Baton Rouge, LA | 98,000 | 10,000 | 4,000 | 6,000 | 400 | 400 | SR, SC, FT |
| Peabody Museum of Natural History Yale University New Haven, CT | 97,000 | 12,000 | 14,000 | 8,000 | 100 | display only | |
| Los Angeles County Natural History Museum, Los Angeles, CA | 91,000 | 5,400 | 2,700 | | | 3,000 | FS |
| National Museum of Natural Sciences National Museums of Canada Ottawa, Ontario | 87,000 | 5,100 | 3,000 | 5,000 | 700 | 1,200 | SR |
| Delaware Museum of Natural History Greenville, DE | 70,000 | 4,600 | 3,500 | 35,000 | 200 | 500 | |
| Moore Laboratory of Zoology Occidental College Los Angeles, CA | 66,385 | 1,068 | 189 | 396 | 158 | | SR |
| Department of Biology University of California Los Angeles, CA | 53,600 | 2,832 | 300 | | | 3 | |
| Museum of Natural History University of Kansas Lawrence, KS | 45,000 | 20,000 | 3,450 | 1,800 | <10 | | FS |
| San Diego Natural History Museum San Diego, CA | 40,000 | 2,000 | 50 | 545 | 10 | 10 | P |
| Denver Museum of Natural History Denver, CO | 38,000 | | | 7,100 | 900 | 500 | |
| Section of Ecology & Systematics Langmuir Laboratory, Cornell University, Ithaca, NY *Laboratory of Ornithology | 37,000 | 4,000 | 700 | 1,800 | 400 | 500 | SR* |
| Western Foundation of Vertebrate Zoology, Los Angeles, CA | 35,000 | | | 150,000 | 12,000 | 600 | |
| Florida State Museum University of Florida Gainesville, FL **eggs & nests combined | 13,750 | 7,100 | 52+ | 13,100** | | 50 | SR |

| | | | | | | | |
|---|---|---|---|---|---|---|---|
| Department of Ecology & Evolutionary Biology, University of Arizona Tucson, AZ | 13,400 | 1,000 | | 200 | 50 | 25 | FS, P |
| Burke Memorial Museum University of Washington Seattle, WA | 12,000 | 6,000 | | 2,000 | 300 | 400 | FS |
| Bernice P. Bishop Museum Honolulu, HI | 13,000 | 500 | 800 | 400 | 50 | 300 | |
| Texas Cooperative Wildlife Collection Texas A&M University College Station, TX | 10,900 | 100 | 225 | 140 | | | P, SC |
| University of Wisconsin Zoological Museum, Madison, WI | 8,000 | 1,750 | 20 | 1,435 | 30 | 300 | |
| Department of Biology University of Miami, Coral Gables, FL | 7,000 | 6,000 | 100 | 100 | | 100 | |
| Pierce Brodkorb Collection Department of Zoology, University of Florida, Gainesville, FL | 5,000 | 10,000 | | | | | F |
| Charles R. Conner Zoological Museum Washington State University Pullman, WA | (responded to questionnaire, but size of collection not included) | | | | | | |

Legend for "other" category: C, cleared and stained; F, fossils; FS, flat skins; FT, frozen tissues; P, photos &/or slides; SC, stomach contents; SR, sound recordings; SW, spread wings.

APPENDIX II.—Addresses of organizations.

American Association for State and Local
    History
    172 Second Ave. North, Suite 102
    Nashville, TN 37201
American Association of Museums
    1055 Thomas Jefferson Street, NW
    Washington, D.C. 20007
American Ornithologists' Union
    K. P. Able, Treasurer
    P.O. Box 44
    Berne, NY 12023
Association of Systematics Collections
    Museum of Natural History
    University of Kansas
    Lawrence, KS 66045
British Ornithologists' Union
    c/o The Zoological Society of London
    Regent's Park
    London, NW1 4RY England
Canadian Museums Association
    331 Cooper Street, Suite 400
    Ottawa, Ontario
    K2P 0G5 Canada
Cooper Ornithological Society
    Charles T. Collins, Treasurer
    Department of Biology
    California State University
    Long Beach, CA 90840
International Council of Museums
    Maison de L'Unesco
    1, rue Miollis
    75732 Paris-Cedex 15
    France
Museum Computer Network, Inc.
    E.C.C., Bldg. 26
    State University of New York at Stony Brook
    Stony Brook, NY 11794
Museum Reference Center
    A & I Building, Room 2235
    Smithsonian Institution
    Washington, D.C. 20560
Natural Sounds Library
    Laboratory of Ornithology
    159 Sapsucker Woods Road
    Ithaca, NY 14850
Nest Record Card Program
    Laboratory of Ornithology
    159 Sapsucker Woods Road
    Ithaca, NY 14850

Ornithological Societies of North America
    P.O. Box 21618
    Columbus, OH 43221
Smithsonian Oceanographic Sorting Center
    Smithsonian Institution
    Washington, D.C. 20560
Society of American Archivists
    330 S. Wells Street, Suite 810
    Chicago, IL 60606
U.S. Fish and Wildlife Service
    Division of Law Enforcement
    (Region 1: CA, HI, ID, NV, OR, WA)
        Lloyd 500 Building, Suite 1490
        500 NE Multnomah Street
        Portland, OR 97232
    (Region 2: AR, NM, OK, TX)
        P.O. Box 329
        Albuquerque, NM 87103
    (Region 3: IL, IN, IA, MI, MN, MO, OH,
        WI)
        P.O. Box 45, Federal Building
        Fort Snelling
        Twin Cities, MN 55111
    (Region 4: AL, AR, FL, GA, KY, LA, MS,
        NC, PR, SC, TN)
        P.O. Box 4839
        Atlanta, GA 30302
    (Region 5: CT, DE, MA, ME, MD, NJ, NH,
        NY, PA, RI, VT, VA, WV)
        P.O. Box "E"
        Newton Corner, MA 02158
    (Region 6: CO, KS, MT, NE, ND, SD, UT,
        WY)
        P.O. Box 25486
        Denver Federal Center
        Denver, CO 80225
    (Region 7: AK)
        P.O. Box 4-2597
        Anchorage, AK 99509
U.S. National Park Service/Division of
    Museum Services
    Harpers Ferry Center
    Harpers Ferry, WV 25425
VIREO
    Academy of Natural Sciences
    19th and The Parkway
    Philadelphia, PA 19103